Sprayed Concrete Technology

SCA/ACI INTERNATIONAL CONFERENCE 1996
Reviewers Committee

Austin S	Dept of Civil and Building Engineering, Loughborough University of Technology	Loughborough, UK
Banthia N	University of British Columbia	Vancouver, Canada
Barton R	Sika Ltd	Welwyn Garden City, UK
Beaupre D	Laval University	Quebec, Canada
Dargan B P	Makers Industrial Ltd (Chairman SCA)	Warrington, UK
Edgington J	Fibre Technology Ltd	Pinxton, UK
Gebler S	CTL	Skokie, USA
Glassgold L	Masonry Resurfacing Construction Co Inc	Baltimore, USA
Gordon K		Kempsey, UK
Lewis R C	Elkem Materials Ltd	High Wycombe, UK
Morgan D R	AGRA Earth & Environmental Ltd	Burnaby, Canada
Quarton P	Connaught Group Ltd	Gloucester, UK
Rimoldi A	Concrete Repairs Ltd	Mitcham, UK
Robins P	Dept of Civil and Building Engineering, Loughborough University of Technology	Loughborough, UK
Taylor G	GT Concrete Training	Warlingham, UK
Woolley G R	Dept of Civil Engineering, University of Leeds	Barkston Ash, nr Tadcaster, UK

Sprayed Concrete Technology

The Proceedings of the ACI/SCA International Conference on Sprayed Concrete/Shotcrete

"Sprayed Concrete Technology for the 21st Century" held at Edinburgh University from 10th to 11th September 1996 organised jointly by the American Concrete Institute and Sprayed Concrete Association

EDITED BY

S. A. Austin

*Department of Civil & Building Engineering
Loughborough University, Loughborough, UK.*

London · Weinheim · New York · Tokyo · Melbourne · Madras

Published by Chapman & Hall, 2–6 Boundary Row, London SE1 8HN, UK

Chapman & Hall, 2–6 Boundary Row, London SE1 8HN, UK

Chapman & Hall GmbH, Pappelallee 3, 69469 Weinheim, Germany

Chapman & Hall USA, 115 Fifth Avenue, New York, NY 10003, USA

Chapman & Hall Japan, ITP-Japan, Kyowa Building, 3F, 2-2-1 Hirakawacho, Chiyoda-ku, Tokyo 102, Japan

Chapman & Hall Australia, 102 Dodds Street, South Melbourne, Victoria 3205, Australia

Chapman & Hall India, R. Seshadri, 32 Second Main Road, CIT East, Madras 600 035, India

First edition 1996

© S.A. Austin 1996

Printed in Great Britain by St Edmundsbury Press, Bury St Edmunds, Suffolk

ISBN 0 419 22270 7

Apart from any fair dealing for the purposes of research or private study, or criticism or review, as permitted under the UK Copyright Designs and Patents Act, 1988, this publication may not be reproduced, stored, or transmitted, in any form or by any means, without the prior permission in writing of the publishers, or in the case of reprographic reproduction only in accordance with the terms of the licences issued by the Copyright Licensing Agency in the UK, or in accordance with the terms of licences issued by the appropriate Reproduction Rights Organization outside the UK. Enquiries concerning reproduction outside the terms stated here should be sent to the publishers at the London address printed on this page.

The publisher makes no representation, express or implied, with regard to the accuracy of the information contained in this book and cannot accept any legal responsibility or liability for any errors or omissions that may be made.

A catalogue record for this book is available from the British Library

Publishers Note This book has been prepared from camera ready copy provided by the individual contributors in order to make the book available for the Conference

∞ Printed on permanent acid-free text paper, manufactured in accordance with ANSI/NISO Z39.48-1992 (Permanence of Paper).

Contents

Preface		ix
Acknowledgements		xi
	PART ONE **ADMIXTURES**	**1**
1	**Powdered air-entraining admixture in dry-mix shotcrete** D. Beaupre, A. Lamontagne, M. Pigeon and J.-F. Dufour	3
2	**Properties of dry sprayed concrete containing ordinary portland cement or fly ash-portland cement** J.G. Cabrera and G.R. Woolley	8
3	**New admixtures for high performance shotcrete** K.F. Garshol	26
4	**Advanced experiences with high performance alkalifree non-toxic powder accelerator for all shotcrete systems** D. Mai	39
	PART TWO **APPLICATIONS**	**47**
5	**Sprayed concrete in conjunction with electro chemical protection of concrete** K. Dykes	49
6	**A dome for your home (or office) - innovative uses of steel fibre reinforced shotcrete** L.E. Hackman and M.B. Farrell	56
7	**Low cost laminated shotcrete marine structures** M.E. Iorns	63
8	**Recent developments in materials and methods for sprayed concrete linings in soft rock shafts and tunnels** R. Manning	70
9	**Sprayed concrete for the strengthening of masonry arch bridges** C.H. Peaston, B.S. Choo and N.G. Gong	81
10	**Application of thick and heavily reinforced shotcrete** J. Warner	89

vi *Contents*

	PART THREE FIBRES	**97**
11	**Evolution of strength and toughness in steel fiber reinforced shotcrete** A.D. Figueiredo and P.R.L. Helene	99
12	**Flexural strength modelling of steel fibre reinforced sprayed concrete** P.J. Robins, S.A. Austin and P.A. Jones	107
	PART FOUR MATERIALS	**115**
13	**Dry shotcrete with rapid hardening cement and humid aggregates** Th. Eisenhut and H. Budelmann	117
14	**Polyolefin fibre reinforced wet-mix shotcrete** D.R. Morgan and L.D. Rich	127
15	**Sprayed fibre concrete: the way forward in civil engineering** M.J. Scott	139
	PART FIVE REPAIR	**147**
16	**The repair of Gorey Jetty in the state of Jersey** K. Armstrong and P. Quarton	149
17	**Wet process sprayed concrete technology for repair** S.A. Austin, P.J. Robins, J. Seymour and N.J. Turner	157
18	**Wet sprayed mortars reinforced with flexible metallic fibres for renovation: basic requirements and full-scale experimentation** J.M. Boucheret	166
19	**Sprayed mortar repairs to highway structures** R.B. Dobson	173
20	**Aesthetic flatwork rehabilitation** J.H. Ford	181
21	**Sprayed concrete repairs - their structural effectiveness** G.C. Mays and R.A. Barnes	188
22	**Long term performance of sprayed concrete repair in highway structures** P.S. Mangat and F.J. O'Flaherty	196
	PART SIX SPECIFICATION	**207**
23	**Contractual aspects of testing shotcrete and rockbolts** R.B. Clay and A.P. Takacs	209

24	**Shotcrete standards - an American perspective** I.L. Glassgold	217
25	**Rehabilitation of concrete structures in Germany by sprayed concrete** G. Ruffert	226
26	**Training and certification scheme for sprayed concrete nozzlemen in the UK** G.R. Woolley and C. Barrett	231

PART SEVEN
SPRAY PROCESS   **241**

27	**Particle kinematics in dry-mix shotcrete - research in progress** H.S. Armelin, N. Banthia and D.R. Morgan	243
28	**Fundamentals of wet-mix shotcrete** D. Beaupre and S. Mindess	252

PART EIGHT
TEST METHODS AND PERFORMANCE   **263**

29	**Evaluation and testing of sprayed concrete repair materials** P. Lambert and C.R. Ecob	265
30	**Evaluation of new and old technology in dry process shotcrete** R.J. Liptak and D.J. Pinelle	270
31	**Strengthening of concrete bridges using reinforced sprayed concrete** D. Pham-Thanh, S.R. Rigden, E. Burley and P. Quarton	280

Author index   297

Keyword index   299

Preface

This book presents the proceedings of the American Concrete Institute/Sprayed Concrete Association Conference on Sprayed Concrete held in Edinburgh on 10-11th September 1996. This was the first joint venture between the two organisations, although both have been promoting sprayed concrete (shotcrete) for many years.

The ACI first established its standing committee on this material in 1942 – Committee 805 on Pneumatically Placed Mortar; this was reconstituted in 1960 as Committee 506 on Shotcrete to revise the ACI 'Recommended practice for shotcreting', and has become one of the most authoritative sources of technical information and standards for sprayed concrete.

The SCA, which is the British trade association, was formed in 1976 under the name the Association of Gunite Contractors and later renamed itself the Sprayed Concrete Association to reflect current European terminology. It has held national and international seminars and conferences in the UK since 1979 and is currently supporting European standardisation through EFNARC – the European Federation of National Associations of Specialist Contractors.

The advent of European harmonisation was one of the factors that led the SCA to approach the ACI about a joint conference in 1994. The aim was to provide a platform for an open exchange of information, views and experiences between practising engineers on both sides of the Atlantic. This has been achieved with papers from North and South America and various European countries. It was also expected that the event would contain a significant emphasis on repair applications, as this is a dominant issue in the countries concerned, in contrast to Scandinavia where tunnelling is the main activity. However, a glance through the index will reveal that the agenda was by no means dominated by repair, but reflected a broad spectrum of interests including specification, test methods, admixtures, fibre reinforcement, materials, the spraying process and performance.

The development of sprayed concrete will depend on how the challenges to produce cheaper, higher performance, more durable concrete are met. Consistent high quality is clearly a key issue and appropriate developments in materials technology, equipment and design methods must deliver the high performance that will be demanded of construction materials in the next decade. Some of the papers reflect that there have been significant changes in the materials used in sprayed concrete over the last ten years. Silica fume has become a common addition to increase adhesion, strength and impermeability, and admixtures have extended their influence from simple accelerators to include water-reducers, super-plasticisers, air-entraining agents, hydration stabilisers and latex polymers. Fibre reinforcement (mainly steel but also polypropylene) has also become a common addition in both dry and wet processes to improve toughness and impact resistance, but there continues to be substantial debate over appropriate definitions and test methods. Special pre-bagged combinations of these materials are available to give high performance in certain applications, including repair.

Progress in equipment has been less marked. Most contractors continue to use conventional rotating barrel/bowl guns for the dry process and worm or piston pumps for wet mix. Steady improvements have been made in recent years, including the introduction of flexible machines that can spray dry or wet. In some countries there has been a big swing towards the wet process, partly because of the better control over mix proportions (particularly the water/cement ratio). In the USA the two techniques have a roughly even share, whereas in the UK the proportion is low but increasing, whilst countries like Germany are still predominantly oriented to the dry process. These differences partly reflect the functional emphasis, but also what is perceived to be good practice in the engineering community. Undoubtedly there will be further shifts in the balance as trade barriers are removed and technology is transferred across borders and fields of application. Dry mix proponents need to address the consistency and quality issues associated with manually controlled water addition, whilst the wet mix process range of applications could be extended by improving its stop/start flexibility.

The proceedings suggest that there needs to be more research and development of design methods. In recent years the industry has benefited from the introduction of a range of standard test methods and specifications. These must be regularly updated and augmented by other appropriate documentation. In Europe the Committee for European Normalisation has just established a working group to produce a European standard specification for 1997. However, engineers also need a framework of acceptable design methods to cover the range of sprayed concrete structures, in particular repair and slope stabilization, to help quantify and justify their design proposals.

It is hoped that the text will provide a ready source of reference for consulting, contracting and materials engineers, clients, manufacturers, students and researchers who seek views of current thinking in this area of concrete technology.

Simon Austin
May 1996

Acknowledgments

The production of this text is the result of much effort and co-ordination, as all those who have personal experience of producing an edited work will understand. We would like to thank the Sprayed Concrete Association Organising Committee members for their involvement in the promotion and organisation both of the Conference and of these proceedings. We are also grateful for the assistance of the members of the Technical Review Committee, drawn largely from members of the Sprayed Concrete Association and American Concrete Institute, who have refereed the papers diligently.

Our thanks also go to the authors, mainly busy practising engineers, who have taken the trouble to submit and prepare their papers, and thus share their experiences with the wider industrial and academic communities.

Our final thanks are reserved for Lisa Lawrence, secretary at the Sprayed Concrete Association, who was responsible for much of the detailed organisation of the submission and refereeing of the papers and also the final manuscript preparation. Her initiative, hard work and patience were a very welcome and important contribution.

Simon Austin (Editor)
John Fairley (Secretary to the Sprayed Concrete Association)

PART ONE
ADMIXTURES

1 POWDERED AIR-ENTRAINING ADMIXTURE IN DRY-MIX SHOTCRETE

D. Beaupre, A. Lamontagne and M. Pigeon
Laval University, Sainte-Foy, Canada

J.-F. Dufour
King Packaged Materials Company, Burlington, Canada

Abstract
In Québec, liquid air-entraining admixtures are commonly used to improve the durability of dry-mix shotcrete repairs exposed to severe winter conditions. The liquid air-entraining admixture is mixed with the water that is added at the nozzle during the shooting operation, particularly in order to improve the resistance of shotcrete to scaling due to freezing in the presence of deicer salts. This procedure requires special equipment and may cause problems in the field. This paper presents the results of a study undertaken to verify the efficiency of powder air-entraining admixtures to entrain an adequate air void system in dry-mix shotcrete. Eleven mixtures were shot, using three powder and one liquid air-entraining admixtures. The test program included compressive strength, deicer salt scaling resistance, spacing factor and the volume of permeable voids tests. The results obtained show that powder air-entraining admixtures can improve both the quality of the air void system (i.e. lower the spacing factor), as well as the resistance to scaling due to freezing and thawing in presence of deicer salts.
Keywords: compressive strength, dry-mix shotcrete, powder air-entraining admixture, scaling, spacing factor.

1 Introduction

In North America, particularly in the province of Québec, many structures exposed to severe winter conditions need to be repaired. The main causes of deterioration are the corrosion of the reinforcing steel and the action of freezing and thawing cycles in the presence of deicer salts. These repairs are mostly superficial with a thickness that is usually of the order of 50 to 100 mm. In the last five years, dry-mix shotcrete has been used extensively for this type of thin repairs.

Two conditions must be fulfilled in order to obtain durable repairs: the repair material must be durable and the bond between the new material and the old concrete must also be durable. To improve the durability of the shotcrete to freezing and thawing cycles in the presence of deicer salts, the following technique has been developed: a liquid air-entraining admixture is mixed with the water added at the nozzle. This has been observed to enhance very significantly the durability of the repair material. To obtain a durable bond, proper surface preparation and construction practices are required [1]. Usually, shotcrete develops good bonding characteristics with the old concrete [2].

Many studies have shown the beneficial effect of using liquid air-entraining admixtures in dry-mix shotcrete [2, 3, 4]. However, several inconveniences are associated with the use of this method: it requires additional equipment (tank and pump), and the dosage of the air-entraining admixture at the job site can be a source of error. After a few years of practical use, it has been found that it is difficult to verify if the air-entraining admixture and the water are effectively mixed in the right proportions by the contractor (usually 20 ml of admixture per liter of water). It has also been found that the measurement of the fresh air content (by shooting directly into the base of the airmeter) does not yield significant results for dry-mix shotcrete [5, 6]. A better way to control the dosage of the air-entraining admixture would be very useful to insure a better quality control of the shotcrete.

Another possible drawback of mixing the air-entraining admixture with the water added at the nozzle is that the quality of the air void system may be influenced when the nozzleman reduces the amount of water in the mixture. Especially when the mixture is shot in an overhead position, the amount of water is reduced by the nozzleman to obtain the properties needed for the mixture to stay in place. The amount of admixture is thus reduced and the effect of this reduction in the admixture content is not precisely known. Also, it is common practice to humidify the surface before shooting with the water used for shooting. Because the water contains an air-entraining admixture, this could affect the bond between the old and newly applied shotcrete.

One of the possibilities of improving the quality control of the shotcrete operations is to replace the liquid air-entraining admixture with a powder admixture that can be simply mixed at the plant with the other dry materials. The aim of the tests described in this paper was to verify if such powder air-entraining admixtures can be used to entrain a proper air void system in dry-mix shotcrete even if the mixing time is very short.

2 Research Program

This project represents a preliminary step to check the performance of different powder air-entraining admixtures for the production of an adequate air void system in dry-mix shotcrete. Eleven mixtures were shot and tested. The type and the dosage of the air-entraining admixtures were the main variables.

3 Materials, mixtures and operating procedures

The same mixture proportions were used for all shotcretes: 22% of CSA type 30 cement (ASTM type III), 60% fine aggregate, and 18% 10 mm crushed limestone. The grading of these aggregates (Table 1) is very close to the limits proposed by the ACI 506 committee. All mixtures also contained 1 kg of polypropylene fibers as specified by the Québec ministry of transportation for repair work.

The eleven mixtures included one reference mixture without any admixture, one reference mixture with a liquid air-entraining admixture, and nine others made with three different powder admixtures (A, B and C) at three different dosages (0.4%, 1.2% and 4.0% of the cement weight). Admixtures A and C contain both a sodium soap of rosine and neutralized fatty acids while the composition of B is not known. The code name of the mixtures describe the admixture used and the dosage (e.g.: A-0.4). The dosages selected cover a wide range since, at the beginning of the project, there was no indication of the results to be expected.

The shotcrete equipment used was a double chamber gun mounted with a batch mixer. The compressor used has a capacity of 20 m^3/s (700 cfm). The nozzle used was a "long nozzle", i.e. a nozzle with the water ring placed 3 meters back from the end of the hose.

Table 1. Gradation for combined aggregates

Sieve size, U.S. standard square mesh	Grading (% passing)	ACI Gradation No.2
1/2" (12.7 mm)	100.0	100
3/8" (9.52 mm)	98.6	90 - 100
4 (4.76 mm)	74.5	70 - 85
8 (2.38 mm)	63.0	50 - 70
16 (1.19 mm)	51.5	35 - 55
30 (0.59 mm)	36.1	20 - 35
50 (0.297 mm)	14.2	8 - 20
100 (0.149 mm)	2.8	2 - 10

For every mixture, two test panels (500 mm x 500 mm x 150 mm) were shot in a vertical position. The surface to be tested for scaling resistance was finished with a wooden trowel. All panels were water cured for 7 days. The cores and the other specimens required for the different tests (compressive strength, characteristics of the air void system (ASTM C457), volume of permeable voids (ASTM C642), and deicer salt scaling resistance (ASTM C672)) were extracted from these test panels. The compressive strength tests were performed on 75 mm diameter and 125 mm long cores.

4 Test results and discussion

The complete test results are presented in Table 2. These results include the compressive strength at 28 days (7 days of curing followed by 21 days of laboratory drying), the volume of permeable voids, the air content and the spacing factor measured on the hardened concrete, and the mass of residues after 50 cycles of freezing and thawing in the presence of a deicer salt solution.

4.1 Compressive strength

As can be seen in Table 2, the compressive strength of the reference mixture containing no air-entraining admixture (52 MPa) is normal for this type of shotcrete and higher than that of all the other mixtures. The reference mixture with the liquid admixture has a compressive strength of 33 MPa. All shotcretes containing a powder air-entraining admixture have a compressive strength close to this value, except that containing the highest dosage of admixture A (A-4,0). Although setting times were not measured, set retardation was observed on this mixture, which indicates that an extremely high dosage of the admixture may influence the hydration process.

4.2 Permeable voids

The volume of permeable voids (Table 2) varies between 12,3% and 14,9% for all shotcretes, except for A-4,0 (19,5%). This confirms the previous conclusion that an extremely high dosage of the admixture may influence the hydration process.

4.3 Air void characteristics

The values of the air void spacing factor in Table 2 confirm that both liquid and powder admixtures can be used to obtain an adequate air void system in dry-mix shotcrete. The non-air-entrained mixture has the highest spacing factor (415 µm), whereas the average value for all the other mixtures is 165 µm. These results also show that a dosage of 0,4% for powder air-entraining admixtures was sufficient in all cases. They further confirm that there is little correlation between the air content and the spacing factor in dry-mix shotcrete.

Table 2. Test results

Mixture	Compressive Strength (MPa)	Permeable Voids (%)	Air Content (%)	Spacing Factor (µm)	Scaling Residues (kg/m^2)
A-0.4	40	14.9	8.0	169	0.6
A-1.2	26	14.7	6.2	221	2.5
A-4.0	12	19.5	8.6	234	3.6
B-0.4	36	12.6	5.3	164	1.0
B-1.2	38	12.6	6.2	101	0.1
B-4.0	28	14.1	9.1	64	--
C-0.4	37	12.3	5.1	146	0.6
C-1.2	31	12.8	8.0	173	0.3
C-4.0	30	14.1	8.7	177	0.3
Reference[1]	52	14.0	4.7	415	8.8
Liquid AEA[2]	33	13.8	6.3	185	0.3

1 Mixture containing no air-entraining admixture
2 Mixture with a liquid air-entraining admixture

4.4 Salt scaling resistance

The mass of scaling residues after 50 freezing and thawing cycles in the presence of a deicer salt solution (3,0% NaCl) is a very good indication of the resistance to scaling. It is generally considered that the scaling resistance is adequate if the mass of scaling residues after 50 cycles does not exceed 1,0 kg/m^2.

The values in Table 2 vary between 0,11 kg/m^2 and 8,81 kg/m^2, but only 3 values exceed the 1,0 kg/m^2 limit: that corresponding to the non-air-entrained reference mixture

(8,81 kg/m^2), and those corresponding to the mixtures made with the high dosages of admixture A (2,54 kg/m^2 and 3,60 kg/m^2). These results show that powder air-entraining admixtures can improve very significantly the scaling resistance of dry-mix shotcrete if the dosage is appropriate. From these results, a dosage in powder air-entraining admixtures of 0,4 % of the cement weight appears to be adequate.

5 Conclusion

The results presented in this paper show that powder air-entraining admixtures in dry-mix shotcrete can be used successfully to reduce the spacing factor and thus increase the deicer salt scaling resistance. The dosage of these materials is a very important parameter because they may affect the hydration process, which could explain the low compressive strength and the reduced improvement in salt scaling resistance observed at a very high dosage (even if the spacing factor was found to be adequate).

6 Acknowledgment

The authors are grateful to the Natural Sciences and Engineering Research Council (NSERC) of Canada for its financial support for this project through the "Chaire Industrielle sur le béton projeté et les réparations en béton" (Industrial Chair on Shotcrete and Concrete Repairs).

7 References

1. Pigeon, M., Saucier, F. (1992) Durability of Repaired Concrete Structures, Advances in Concrete Technology, pp. 741-773.

2. Talbot, C., Pigeon, M., Beaupré, D., Morgan, D.R. (1994) Long Term Bonding of Shotcrete, *ACI Materials Journal*, Vol. 91, No. 6, pp. 560-566.

3. Beaupré, D., Talbot, C., Gendreau, M., Pigeon, M., Morgan, D.R. (1994) Deicer Salt Scaling Resistance of Dry and Wet Process Shotcrete, *ACI Materials Journal*, Vol. 91, No. 5, pp. 487-501.

4. Lamontagne, A., Pigeon, M., Pleau, R., Beaupré, D. (1996) Use of Air Entraining Admixture in Dry-Mix Shotcrete, *ACI Materials Journal*, Vol. 93, No. 1, pp. 69-74.

5. Morgan, D.R. (1991) Freeze Thaw Durability of Steel and Polypropylene Reinforced Shotcretes: a Review", Second CANMET/ACI International Conference on Durability of Concrete, Montréal, Canada, pp.901-918.

6. Beaupré, D., Lamontagne, A. (1995) The Effect of Polypropylene Fibers and Aggregate Grading on the Properties of Air-Entrained Dry-Mix Shotcrete, Fiber Reinforced Concrete: Modern Developments, Second University-Industry Workshop on Fiber Reinforced Concrete, Toronto, Canada, 26-29 march, pp. 251-161.

7. ACI Committee 506 (1990) Guide to Shotcrete, *American Concrete Institute*, Detroit, Michigan, 41 p.

2 PROPERTIES OF DRY SPRAYED CONCRETE CONTAINING ORDINARY PORTLAND CEMENT OR FLY ASH-PORTLAND CEMENT

J.G. Cabrera and G.R. Woolley
Civil Engineering Materials Unit, Dept of Civil Engineering, University of Leeds, UK

Abstract
The considerable quantity of rebound generated by the Dry Mix process of Sprayed Concrete affects the expectation in-situ and is of commercial and engineering concern. It has been shown that the very high cement contents found in hardened sprayed concrete provide potential for alkali-silica reaction and incipient cracking through high heat of hydration. There is also clear evidence of weaker lenses of material in hardened concrete.

The Paper presents an analysis of performance related properties of cores obtained from dry mix process sprayed concrete during the strengthening of a natural draught cooling tower shell. Density, porosity and permeability of the concrete is analysed and related to the durability of the material. The cement content of the concrete cores is obtained by a rapid thermogravimetric method. Enrichment of the concrete in terms of cement content results in a very high value of sodium oxide equivalent which can lead to problems related to alkali aggregate reaction.

In a preliminary study the use of fly ash as a proportion of the cementitious content to act as a viscosity modifier is described. Reduction in rebound and a more compact concrete in-situ was observed as well as a reduction in sodium oxide equivalent together with commercial benefits to users of sprayed concrete.

Keywords: Sprayed concrete, fly ash, strength, permeability, density, composition, Alkali-silica reaction, rebound.

1 Introduction

"Sprayed Concrete" known as "Shotcrete" in the USA, has increased in popularity due to its versatility for repairing damaged concrete and for strengthening concrete structures. It is also a method widely used for rock support, lining of tunnels, road cuttings and embankment slopes (1, 2).

A recognised drawback of sprayed concrete arises from the method of placement which produces variable amounts of loss of the constituents of the sprayed material by rebound against the surface being sprayed. The loss of material varies with parameters like angle of spray, nature of the surface being sprayed, pressure of spraying, distance from the spraying front and the composition of the concrete itself. Reports on the amount of rebound vary from 0% to 40% depending on the operational variables listed and on the nature of the constituents of the concrete. It is also recognised that dry mix spraying results in larger percentages of rebound than wet mix spraying (3, 4, 5). Apart from the economic loss arising from rebound losses, the major problem is that the rebound material is relatively rich in sand and aggregate when compared with the original mix; this results in a mix in place which is rich in cement and thus can be susceptible to cracking induced by drying and by temperature changes. A rich cement mix also contains excessive amounts of alkali and can potentially activate aggregates containing amorphous and microcrystaline silica to produce alkali-silica reaction products (6,7). For these reasons researchers and specialist firms have produced a variety of "sprayed concrete" compositions. As well as the normal constituents of cement, sand and aggregates, these compositions include: fibres, plasticisers, accelerators, retarders, cohesive agents and other chemicals, with the result that these materials are far more expensive. A major disadvantage is that the user does not know the composition of the material being used and most importantly, neither the user nor the producer knows what, if any, the long term effects are when using a combination of organic and inorganic compounds with Portland cement.

This paper has been prepared with two objectives in mind. Firstly, it presents data that quantifies the composition of a sprayed concrete by measuring cement content in cores drilled from the sprayed concrete used for strengthening a natural draught cooling tower at the Didcot Power Station in the UK, and by measuring cement content in the rebound material and in the original mix just before spraying. Secondly, it presents preliminary data showing the influence of fly ash-cement on the composition of sprayed concrete. This part of the study was carried out at CITB (Construction Industry Training Board) during an experimental programme of different sprayed concrete compositions.

2 Mix Design for the Didcot Strengthening Project

The aggregates used were predominately Jurassic oolitic and shelly limestone with minor proportions of more angular flint and sandstone with a trace of ironstone. Petrographic examination of the material confirmed these descriptions. Potential reactivity with alkalis using the ASTM standard test method C289-81 (chemical method) (8) were completed and the materials were considered to be non-reactive. The alkali content of the aggregates was less than 0.05% sodium oxide equivalent. Figure 1 gives a typical grading curve for the aggregate.

Careful consideration was given to the alkali content of cements. The cement selected for preparation of the sprayed concrete mix was obtained from the Chinnor Works of the Rugby Portland Cement Company in the UK. Table 1 gives the average chemical and mineralogical composition of this cement.

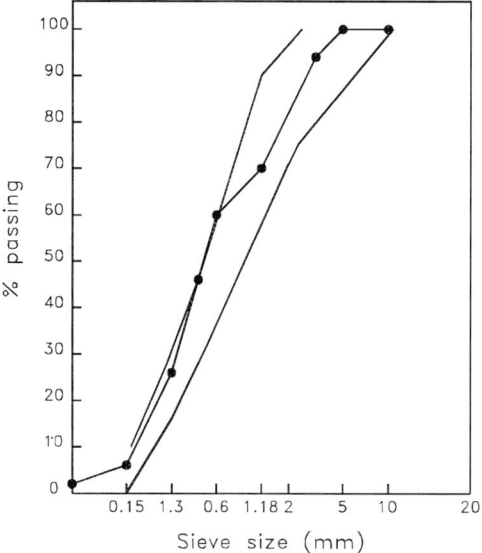

Figure 1. Particle size distribution of the aggregate for the sprayed concrete used in the strengthening of Didcot Cooling Tower.

The sprayed concrete mix designed by the contractor consisted of 1 part by weight of cement and 3.5 parts by weight of sand. Assuming that the total weight of sand plus cement was 2200 kg/m^3 (489 kg of cement and 1711 kg of sand) and that the water per m^3 was 171 kg, the wet density of the mix was 2371 kg/m^3.

The calculated composition, using this mix, of the sprayed concrete prior to placement and after placement is given in Table 2.

Table 1. Chemical and mineralogical composition of the ordinary Portland cement (opc) used.

Oxide	Percentage
SiO_2	21.80
Al_2O_3	3.80
Fe_2O_3	1.50
CaO	66.60
MgO	1.10
Na_2O	0.11
K_2O	0.47
SO_3	3.20
LOI	1.00

Mineral	
C_3S	61.0
C_2S	17.0
C_3A	7.7
C_4AF	4.5

Table 2. Calculated sprayed concrete mix composition before and after placement.

Constituent	Mix before placement kg/m^3	Loss by rebound %	Mix in place kg/m^3
Cement (opc)	489	15	416
Sand 0.5 mm	1551	27	1117
Sand 25 mm	160	63	59
Water	171	17	142
W/C	0.35		0.34

The estimated composition of the mix in place is based on the assumption that the loss of material by rebound is as shown in Table 2. The contractor based his assumption on experience, however, there was no evidence that the mix in place would have the assumed composition. Furthermore, it seemed unrealistic to propose a mix in place with a density of only 1734 kg/m^3. For this reason the contractor used the wet density of the mix prior to placement to calculate a theoretical cement content of the mix in place and from this he obtained the Na_2O equivalent.

The contractor estimated the cement content and Na_2O equivalent of the mix in place as follows:

Cement kg/m^3	2371 x 23.99	=	569
Na$_2$O eq. kg/m^3	Na$_2$O + 0.6 KO$_2$	=	0.39
	Standard Deviation (SD)	=	0.04
	0.39 + 2(SD)	=	0.47
	0.47 x 569	=	2.67

The Na$_2$O eq. is below the limit of 3 kg/m^3 recommended in the Concrete Society Technical Report (9).

The Na$_2$O eq. calculated by the contractor is correct only in that the assumed total weight of constituents is equal to the density of the sprayed concrete mix. This aspect will be discussed later in the light of the results obtained from the cores extracted from the mix in place.

3 Site Control of Mix Composition

Composition control checks on the sprayed concrete were made through the spraying of a series of sample panels. These panels, approximately 1 m^2 comprised a backing sheet set in the upright position on the platform, onto which 100 mm thick concrete was sprayed. The sample panel was then coated with a curing membrane. After 24 hours, cores were cut for curing in accordance with BS 1881 (10) and subsequent compressive strength testing at 7 and 28 days.

Sample checks were undertaken on the amount of rebound produced through the process. Clean scaffold areas were carefully monitored and rebound, after completion of these areas, was recovered. Simple arithmetic checks confirmed the average rate of rebound as a percentage of total application.

To determine permeability and porosity of the sprayed concrete, cores taken from the sample panels were tested using the Leeds Cell developed by Cabrera and Lynsdale (11). Carbonation of cores was checked by phenolthalein indicator.

To establish the actual sodium equivalent of the sprayed concrete applied in the mantle, a sequence of checks were set up to monitor materials, application and rebound. Samples of the dry sand/cement mix, rebound and in-situ sprayed concrete were taken in a sequence designed to be time related through the check. Analysis of the samples by thermogravimetric techniques allowed the determination of cement content and thus Na$_2$O eq. Cores from the mix in place were also obtained to analyse any possible variation of cement content within the depth of the sprayed concrete layer.

4 Results of the Didcot Strengthening Project

A statistical summary of the 7 day and 28 day equivalent cube compressive strength results of samples from the total sprayed concrete application of 21,126 m² is presented in Table 3. A histogram of 28 day compressive strength distribution is shown in Figure 2.

Table 3. Summary of strength results. Didcot sprayed concrete.

Age (days)	Mean equivalent compressive* strength (MPa)	Standard deviation	Number of samples
7	54.07	8.02	64
28	65.10	8.27	135

*Estimated in-situ cube strength determined from 94 mm diameter cores.

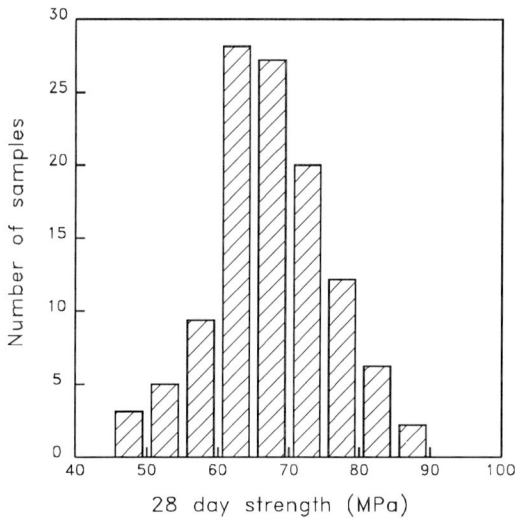

Figure 2. Frequency distribution of the cube compressive strength values of the mix used in Didcot.

Monitoring of rebound material during the field work gave an average rebound value of 21.3% with a standard deviation of 2.56. This value is within the values quoted for rebound of sprayed concrete in vertical walls.

The wet density (D_w) values obtained from the same specimens used to determine compressive strength gave a mean value of 2.322 g/cm^3. The moisture content of the sand was 6% and the moisture content of the cement was on average 10%. Using these moisture values the corrected value for dry density (D_d) was:

$$D_d = 2.322 - (2.322 \times 0.1 \times 0.32 + 2.32 \times 0.68 \times 0.06)$$
$$D_d = 2.153 \text{ g/cm}^3$$

The determination of cement content on samples before placement (dry mix), rebound material and in place mix (cores) was carried out using a thermogravimetric method. This is a rapid method which requires knowledge of the nature of the aggregate and the age of the sample. It is based on the measurement of the amount of bound water, the calcium hydroxide generated and the decomposition of the calcium carbonate. An example of the output of the thermobalance is shown in Figure 3. The accuracy of this method is clearly shown from the values measured using the dry mix for which the composition was known in advance.

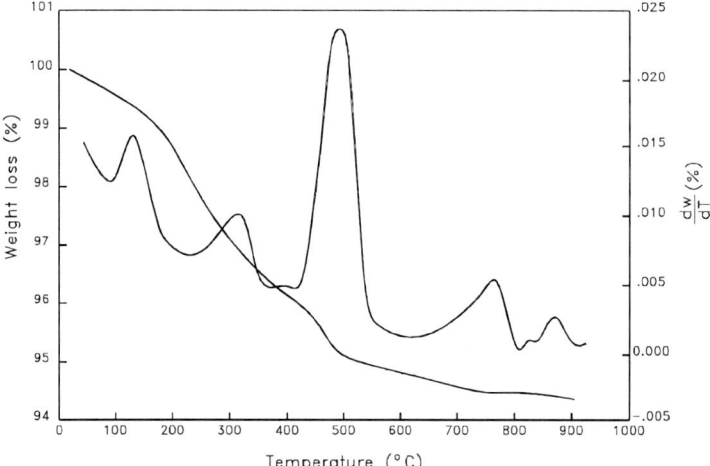

Figure 3. Thermogravimetric result showing cumulative and derivative curves. Structural water loss $\leq$ 400°C. Ca(OH)$_2$ decomposition between 400°C and 540°C. CaCO$_3$ decomposition between 630°C and 800°C.

The final results were corrected using the percent difference between the measurements of the thermobalance and the known mix composition. Table 4 shows these results.

Table 4. Composition of the mixes obtained by thermogravimetric analysis.

Sample No.	Mix type	Cement content	C:A ratio	Corrected C:A ratio	Mix composition C%	A%
14	Core	31.16	1:2.21	1:2.27	32	69
	Rebound	19.10	1:4.24	1:4.35	19	81
	Dry Mix	22.63	1:3.41	1:3.50	22	78
15	Core	39.09	1:1.56	1:1.40	42	58
	Rebound	13.39	1:6.47	1:5.80	15	85
	Dry Mix	20.39	1:3.90	1:3.50	22	78
16	Core	33.29	1:2.00	1:1.37	42	58
	Rebound	15.39	1:5.49	1:3.76	21	79
	Dry Mix	16.32	1:5.10	1:3.50	22	78

The difference in mix composition between the mix in place (cores) and the rebound material is of the same order reported by Hill (6). These results are a clear indication that the mix in place has a very high cement content. The implications of these changes will be discussed in the light of the changes in density and the values of permeability also measured during this study.

The dry density values were measured using the vacuum saturation method and were compared with calculated values obtained using the values of Relative Density (D_r) calculated using the method of Cabrera (12). This method uses the following relations:

$$D_r = \frac{100}{UC/D_{ruc} + HC/D_{rhc} + A/D_r}$$

Where:

D_r = relative density of the mix
UC = unhydrated cement, % by weight
D_{ruc} = relative density of unhydrated cement
HC = hydrated cement, % of weight
D_{rhc} = relative density of hydrated cement
A = aggregate, % by weight
D_{ra} = relative density of aggregate

and:

$$P = (1 - D_d/D_r)\,100$$

Where:

P = porosity of mix, %
D_d = dry density

Direct porosity measurements were also made using the vacuum saturation method (13).

The results are presented in Table 5. This table also includes the calculated values of cement and the <u>real</u> sodium oxide equivalent.

Table 5. Results of relative density (D_r), porosity (P), dry density (D_d), cement content (C) and sodium oxide equivalent.

Core No.	D_r	P (%)	D_d (g/cm^3)	C (%)	C (kg/m^3)	Na$_2$O Eq (kg/m^3)	P_o m^{2}x10^{-15}
14	2.503	10	2.253	32	721	3.39	1.87
15	2.464	13	2.143	42	900	4.23	3.02
16	2.464	13	2.143	42	900	4.23	4.18
Mean value	-	12	2.180	39	840	3.95	3.02
Measured in field samples			2.153	22	474*	2.22*	

P_o = oxygen permeability
* Assumed values

Four cores obtained from the sprayed concrete in place were used to assess the possible variation of cement content within the depth of the sprayed concrete layer. Each core of 50 mm thickness was cut into 3 equal small cores of 12 mm thickness and the cement content of each small core determined by the method already discussed. The small cores corresponded to the inner (I), middle (M) and outer (O) portion of the sprayed concrete layer. The results are presented in Table 6.

Table 6. Variation of cement content and Na$_2$O eq. within the depth of the Didcot sprayed concrete layer.

Sample No.	Position	Cement content (%)	Mix composition C:A	Cement kg/m^3	Na$_2$O eq. kg/m^3
8	I	38	1:1.63	828	3.89
	M	42	1:1.38	916	4.31
	O	36	1:1.78	785	3.69
	Average	39	1:1.56	843	3.96
12	I	37	1:1.79	807	3.79
	M	38	1:1.63	828	3.89
	O	32	1:2.13	698	3.28
	Average	36	1:1.78	778	3.65
18	I	41	1:1.44	894	4.20
	M	34	1:1.94	741	3.48
	O	34	1:1.94	741	3.48
	Average	36	1:1.78	792	3.72
19	I	45	1:1.22	981	4.61
	M	34	1:1.94	741	3.48
	O	30	1:2.33	654	3.07
	Average	36	1:1.78	792	3.72

5 Discussion

From the results it is clear that the sprayed concrete mix used had more than adequate strength; this is normally the case with mixes rich in cement content.

The parameters related to the performance and durability of sprayed concrete measured in this study indicate that the mix in place exhibited low porosity but unusually high permeability. A good concrete will easily reach an intrinsic permeability of 1×10^{-16} m^2, however, the sprayed concrete mix gave values higher by more than one order of magnitude. This is probably due to the presence of large voids which arise from the placement procedure. The compaction of the mix is dependent on the skill of the nozzle operator and is very susceptible to small changes in composition of the mix and therefore, there is a higher probability of producing pockets of loose mix and generation of capillary voids even when the total porosity is low.

A major aspect of concern relates to the richness of the mix in place. Significantly higher than estimated Na$_2$O eq. were found. Cement profiles given in Table 6 are variable, but clearly show a higher cement content near the surface of the concrete being sprayed. Variability of results may, to some extent, be explained by the angle of spray nozzle at placing. The values of Na$_2$O eq. found might in time lead to the generation of alkali-silica reaction products with consequent deterioration of concrete if aggregates used contain silica bearing components. Therefore, specifications for aggregates and tests to control their quality is an important aspect of quality control procedure when producing sprayed concrete.

The high values of cement content found in the inner part of the sprayed concrete layer may lead to excessive heat of hydration. Large differences in cement content between the concrete being repaired or protected and the sprayed concrete may also lead to debonding of the sprayed concrete.

For economic reasons and for the reasons discussed, it seems important and logical that instead of preparing complex mixtures of organic chemical and fibres, research should be carried out exploring the effect of pozzolanic additions which may reduce rebound, and which can certainly reduce the cement content and thus reduce heat of hydration and debonding of the sprayed concrete, but most importantly minimise the risk of alkali aggregate reaction.

The disadvantage when using most pozzolans is that they reduce the early strength of the concrete and/or require higher w/c ratio to maintain satisfactory spraying characteristics. However, fly ash is known to increase the workability of concrete (14) allowing a reduction of w/c ratio and therefore, compensating for the possible loss of early strength. Concrete mixes containing fly ash are very cohesive and therefore rebound of sprayed concrete can be reduced by the plasticising effect of the fly ash.

Heat of hydration is also reduced considerably when using fly ash additions. In the context of the data presented in this paper fly ash can effectively control possible alkali aggregate reactions (15) even when the fly ash contains appreciable quantities of alkali.

5.1 Properties of fly ash-cement sprayed concrete

This part of the investigation involved the spraying of mixes using as a cement a mixture of ordinary Portland cement and fly ash and comparing the performance of these mixes with a sprayed control concrete made only with ordinary Portland cement.

5.2 Experimental Details

Using a mix of 3.5:1 of local quartzitic sand and cementitious material, 750 x 750 x 100 mm deep timber panels were filled with sprayed concrete using the dry mix process. Constituents were combined in a 5/3.5 mixer before being placed in an Aliva-Verio type 246 spraying machine of 2 cubic metres capacity. The dry mix was transported through a 40 mm diameter hose using compressed air delivered by an Atlas Copco XAS 280 compressor.

A control panel was sprayed with the mix design using sand and ordinary Portland cement. Further panels were sprayed with a mixture wherein the cement was replaced with either blended cement to BS 6588 (16) or a blend of 70% ordinary Portland cement with 30% selected fly ash, the latter being combined in the mixer with all other constituents. All proportions were measured by weight and the spraying of all panels was undertaken by the same experienced nozzleman.

On completion of spraying, all panels were coated with a curing agent or wrapped with polythene sheet for the curing process. 100 mm diameter cores were extracted from the panels after 48 hours for curing at 100% RH and testing at 7 and 28 days to establish the estimated in-situ cube strength.

Each panel was erected near vertical on a polythene sheet for the spraying application. All rebound was collected and weighed. Samples of the rebound were placed in sealed plastic bags for later determination of cement content by thermogravimetry. The spraying pressure was varied in order to assess its influence on the properties of the sprayed material. The pressures used were 0.16, 0.20 and 0.24 MPa.

6 Preliminary Results of the Study

6.1 Compressive Strength

Measurement of compressive strength from cores extracted from the panels was made at 7 and 28 days for the spraying pressures measured. These results are shown in Table 7 and also in Figure 4. Although there is a considerable scatter on the values of compressive strength the trend shown in Figure 4 is as has been found by King (17) i.e. increasing air pressure increases the compressive strength of the sprayed concrete independently of the cement type. The differences in compressive strength and cement type are not clear when looking at the results in Figure 4. However, the average values given in Table 7 indicate that fly ash cements gave marginally lower compressive strength at 7 days but statistically equal values at 28 days. This is not surprising since the amount of water required for equal workability is less for the fly ash mixtures and the additional hydration products generated by the pozzolanic reaction increases the volume of the hydrates. Both factors reduce the capillary porosity and therefore compensate for the reduction of the more reactive Portland cement.

Table 7. Estimated cube strength for the three sprayed concretes tested.

Sprayed concrete mix	Compressive strength (MPa)		Spraying pressure (MPa)
	7 days	28 days	
Control (cement + sand)	40.50	45	0.16
	44.79	65	0.20
	-	54	0.24
Blended cement to BS 6588 + sand	35.90	57	0.16
	29.90	41	0.24
Cement + ash (30%) + sand mixer blended	34.50	55	0.16
	32.10	62	0.20
	-	55	0.24

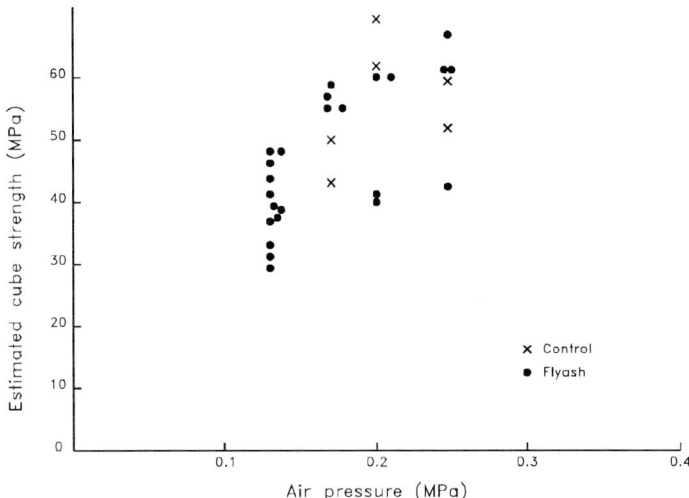

Figure 4. Relationship between air spraying pressure and compressive strength at 28 days.

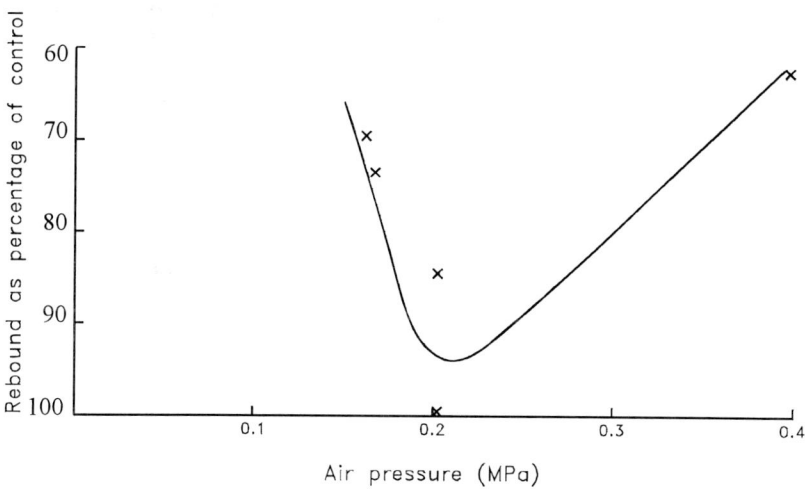

Figure 5. The effect of air pressure on the percentage rebound of the fly ash sprayed concrete mix.

6.2 Effect of air pressure on rebound

The rebound of the fly ash-cement mixes as a function of the air pressure is shown in Figure 5. The rebound as a percentage of that experienced by the control mix was 76% at 0.16 MPa air pressure. This increased to around 90% of the control at 0.2 MPa but when the pressure was increased to 0.4 MPa the amount of rebound decreased considerably. It is suggested that at the higher pressures the velocity at input causes the matrix to form a more congenial thickness into which incoming energy is more easily dispersed.

6.3 Determination of cement content

As for the study described in the first part of this paper the determination of cement content was made using thermogravimetry. The measurements of total structural water from which the cement content was estimated gave the results shown in Table 8. In comparison with the results of Table 4 it is apparent that the loss of aggregate and therefore the enrichment of the fly ash mixes is not of the same magnitude as that of the ordinary Portland cement mix. Observations on site showed that the fly ash mixes were cohesive and 'fattier' than the control mix and therefore were able to hold the aggregate particles which otherwise would have rebounded.

Table 8. Composition of the mixes obtained by thermogravimetric analysis using cores.

Sample No.	Mix	Cement content (%)	C:A ratio by weight
0	Control opc*	23.53	1:3.5
1	OPC/FA (BS 6588)	25.30	1:2.95
2	OPC/FA (Site mix)	20.63	1:3.85
3	OPC/FA (Site mix)	24.75	1:3.0
4	Control**	19.80	1:4.0

* Design composition
** Determination made on rebound material

The rheological changes observed in concretes made with fly ash for normal applications appear to be the same as the changes in the viscosity and rheology of sprayed concrete and therefore as beneficial as with ordinary concrete. This characteristic of fly ash concrete merits further studies so that the parameters for spraying are optimised towards reducing the rebound and the segregation upon impact.

7 Conclusions

From the observations made during the strengthening operations and the results of the tests carried out in field and in the laboratory, the following conclusions are offered.

1. The sprayed concrete mix of design composition of one part of cement to 3.5 parts of sand by weight resulted in a mix in place with a mean composition of less than 1:2. This enrichment gave very high values of compressive strength.

2. The amount of rebound measured in the field gave an average value of 21.3% and the dry density measured in cores gave an average value of 2.153 g/cm^3. This value compares well with the mean value obtained by calculation on the basis of the actual composition of the mix in place which was 2.180 g/cm^3.

3. The values of porosity were relatively low, but the oxygen permeability values were at least one order of magnitude higher than the values for "good concrete". These findings lead to the belief that the mix in place had a considerable volume of capillary pores arising from the uneven pressure of compaction and some microcracking arising from the high cement content. These effects can be minimised by high standards of workmanship and careful control of the in-situ curing procedures.

4. The sodium oxide equivalent was in all cases above the limit recommended as a value to avoid alkali-silica reaction. These findings imply that to avoid harmful effects, it should be imperative to have strict control on the quality of aggregates. Alternatively, it is suggested that blended cements with fly ash are considered which would give reduction in the possibility of alkali-silica reactions.

5. The cement content of the sprayed concrete layer varied substantially from the inside to the outside of the layer. When sprayed concrete is placed in thicker than 50 mm the cement content variations might increase leading to problems of cracking, debonding and alkali aggregate reaction.

6. Use of fly ash to replace a proportion of cement has been shown to offer benefits in the spraying process. The lower than control early strength differences largely disappear with time, indicating that at the more commonly specified 28 days differences are minimal.

7. Reductions in rebound and the observed ability of the fly ash sprayed concrete to compact more readily suggest a denser and more durable product.

8. Reductions in rebound by replacement of a proportion of cement with the less expensive fly ash offer real commercial benefits to users of sprayed concrete.

8 Acknowledgements

The authors wish to record their thanks to The Construction Industry Training Board for allowing the production of the fly ash sprayed concrete panels. To both National Power Plc and to Scottish Power Plc for providing the fly ash and blended cement for the test panels.

9 References

1. Concrete Society. Code of Practice for Sprayed Concrete. The Concrete Society. London, 1980.
2. Robin P J. Physical properties of sprayed concrete in Sprayed Concrete, Properties, Design and Application. Ed Austin S A and Robin P J. Whittles Publishing 1995.
3. Parker H W, Fernandez-Delgado G and Lorig L J. A practical new approach to rebound losses. Shotcrete for Ground Support. American Concrete Institute SP-54, pp 149-187, 1977.
4. Henager C H. The technology and uses of steel fibrous shotcrete. A state-of-the-art report. Report, Batelle Development Corporation. USA, 1977.
5. Ward W H and Hills D L. Sprayed concrete tunnel support requirements and the dry mix process. Shotcrete for Ground Support. American Concrete Institute, SP-54, pp. 475-532, 1977.
6. Hill D L. Site-produced sprayed concrete. Concrete, pp 44-50, No. 12, 1982.
7. Cabrera J G and Woolley G R. Potential for developments in sprayed concrete. Sprayed Concrete European Seminar, Brunel University, UK. 1993.
8. American Society for Testing Materials. ASTM 289-81. Test for Potential reactivity of aggregates (chemical method). ASTM, 1981.
9. Concrete Society. Alkali-silica reaction: minimising the risk of damage to concrete. Technical report No. 30. The Concrete Society, 1987.
10. British Standard BS1881: Part 121:1983. Methods for determination of the compressive strength of concrete cores. The Institution, 1983.
11. Cabrera J G and Lynsdale C J. A new gas permeameter for measuring the permeability of mortar and concrete. Magazine of Concrete Research, Vol.40, No. 144, 1988.
12. Cabrera J G. The Porosity of Concrete. Concrete Research, University of Leeds, June 1985.
13. Cabrera J G and Lynsdale C J. Measurement of chloride permeability in superplasticised opc and pozzolanic cement mortars. Proceedings of International Conference on Measurements and Testing in Civil Engineering. Lyon, France, 1988.
14. Cabrera J G. The use of pulverised fuel ash to produce durable concrete. Institution of Civil Engineers, 19-36, Thomas Telford, London, 1985.
15. Zou Y. Composition of ASR reaction products and their influence on expansion of mortars. Unpublished PhD Dissertation, University of Leeds, UK. 1991.

16. British Standard BS 6588:1993 Specification for Portland pulverised-fuel ash cement. The Institution. 1993.
17. King E. Design for structural applications in Sprayed Concrete Properties, Design and Applications. Eds. Austin S A and Robins P J. Whittles Publishing 1995.

3 NEW ADMIXTURES FOR HIGH PERFORMANCE SHOTCRETE

K.F. Garshol
MBT International Underground Construction Group,
Zurich, Switzerland

Abstract
High performance shotcrete may be understood as:
- the wet mix application method
- reinforcement by steel fibres (SFRS)
- high concrete quality, giving high density, low permeability and good durability
- applied with equipment adapted to the task, allowing low rebound, high safety and excellent quality assurance
- utilisation of latest admixture technology

The application of this technology is often blocked by a too narrow evaluation basis, looking into part costs like materials, equipment investment etc., without a proper calculation of total cost in the structure.

High performance wet mix shotcrete can be produced by using a number of new admixture products. Among these are the liquid alkali-free and non caustic accelerators, the consistency control system, also named "slump killing" and the concrete improver admixture, replacing external spray-on curing membranes.

The advantages of these new admixtures are presented. The issues covered are:
- quality and durability
- time saving and economy, using spread-sheet comparison calculations
- working safety and working environment

The set of arguments in favour of high performance wet mix shotcrete is very strong. The key to a successful acceptance and use of the method, lies in overall economy and durability analysis.
Keywords: Admixtures, economy, fibres, safety, shotcrete, quality.

1 Introduction

High performance wet mix shotcrete may be defined as follows:
- shotcrete applied according to the wet mix method
- normally, reinforced by steel fibres to provide a high ductility
- a high concrete quality on the rock surface, giving high density, low permeability and therefore good durability
- shotcrete applied with high capacity equipment, allowing low rebound, good economy, high safety and excellent quality assurance
- utilisation of latest admixture technology, to meet above requirements and to be able to achieve the best possible working environment

Very often when this technology is being discussed, the question is raised about why such an "expensive" solution shall be chosen. The question is reasonable at first, because the materials cost may increase by as much as a factor of two (steel fibres included), the investment in equipment may increase by a factor of five and sometimes shotcrete is only used for immediate temporary support. The point is that if the evaluation is covering only the above issues, the most important ones are overlooked.

High performance wet mix shotcrete can be produced by using a number of new admixture products. Among these are the liquid alkali-free and non caustic accelerators, the consistency control system, also named "slump killing" and concrete admixture that can replace external spray-on curing membranes.

2 Alkali-free, non caustic accelerators

For high durability and environmental friendly application of shotcrete, a milestone breakthrough has occurred recently. Liquid form accelerators are now available, having a pH in the range of 3 to 5.5 and an alkali content of less than 0.3%. When comparing to the normal aluminate accelerators, having pH about 14 and alkali content of about 20%, the differences and advantages become obvious.

The liquid form accelerators also overcome the working environment problem of available alkali free powder products. Powders produce quite a lot of dust, which is unfavourable, even when the product is non aggressive and non corrosive. Using a liquid, the dosing becomes much more accurate and simple and the overall handling and logistics is simplified.

The liquid accelerators are normally added at a rate of 3 to 10% of the binder weight. At such dosages, the layer thickness in the roof may be up to 300 mm and the strength development is normally between and partly above the J2 and J3 curves of the Austrian shotcrete guidelines. Typically, a compressive strength of 15 to 25 MPa is reached after 24 hours, depending on dosage, temperature and type of cement. A very important feature is: There is no, or only marginal loss of 28 day strength, compared to a zero concrete (without accelerator).

The points to be aware of when considering an alkali-free accelerator, are quite simple:

- check the compatibility of the cement. Simple laboratory tests are available, that will produce a good indication
- the w/c- ratio should be kept below 0.5, preferably below 0.45
- the fresh concrete slump test should be below 20 cm or less than 42 cm spread
- if a hydration control admixture is being used, check the practical dosage range
- stir the stored containers before use and clean the dosage pump once per day
- with accelerator pH on the acid side, check the corrosion protection of the dosing equipment

3 The consistency control system

Consistency control for shotcrete application is a supplement to the use of pure accelerators. The system contains two components. The first component is a special super plasticiser, that is added at the batching plant. The second component is an activator, which is added in the spraying nozzle. A high slump, workable material is created by the plasticiser and this slump is "killed" in the nozzle by the activator, so that a plastic, low slump concrete hits the substrate.

This system may utilise slump killing alone, without set acceleration (allowing more than an hour for surface finishing), or can be combined with quick setting accelerator for thick layers and high early strength. In any case, it is very important that the concrete remains plastic and thixotropic for a few seconds after impact, because this reduces rebound and improves compaction around reinforcement. In particular, if very long steel fibres (40 mm length) are used, a dramatic reduction of fibre rebound is achieved.

The properties of the consistency control system, was the basis for its application for the construction of the bob sleigh track at the Lillehammer Winter Olympics.

3.1 The Hvassum Road Tunnel, Norway

The 3.3 km Hvassum road tunnel is located about 20 km outside of Oslo. The drill and blast excavation through granitic gneiss, is permanently supported by SFRS and fully grouted rock bolts. Shotcrete thickness varies between 75 and 150 mm.

One special feature of this project, was the Client's requirement on toughness. The required I_{30} index, according to ASTM 1018, was as high as 22. This turned out to be difficult, partly due to the high required compressive strength of 40 MPa. The contractor, Veidekke AS, tried longer fibres (40 mm at 0.5 mm thickness). This was not a solution, due to increased fibre rebound. Trials, with more silica fume and other variations in the mix design, could not bring the desired results. The toughness index varied between 13 and 20.

The consistency control system turned out to be the solution. This "slump killing" system retains a short period of plasticity of the shotcrete after impact on the substrate. This improves compaction from subsequent shotcrete impacts and it reduces fibre rebound and improves fibre orientation. The practical result was a fibre rebound reduction from 20 - 25%, down to 10 - 12%. The toughness index I_{30} increased to above 25.

Table 1:

Description	Old mix design	Cons. control system
Portland cement	470 kg	470 kg
Microsilica, densified	8%	8%
Aggregate 0 to 8 mm	1670 kg	1670 kg
Superplasticiser (BNS)	5 kg	
Superplast. for cons. control		8 kg
Plasticiser (lignosulphonate)	3.5 kg	
Steel fibres l=40, d=0.5 mm	50 kg	50 kg
w/c+s	0.40	0.41
Slump	12 to 16 cm	20 to 24 cm
Accelerator	5%	
Activator, polymer mod. silicate		4%
Rebound total	5 to 6%	4 to 5%
Fibre rebound	20 to 25%	10 to 12%
I_{30} index	13 to 20	all > 25
Open time concrete mix	1 hour	more than 2 hours

The shotcrete quantity remaining on the project, after the change to the consistency control-system, amounted to 4000 m^3.

4 Curing of shotcrete

This subject should create bad conscience among most tunnelling professionals. The normal practice is to not cure at all. This situation is caused by the practical problems of constant water spraying, and even the cost and hassle of applying a spray-on membrane, that afterwards must be removed if another layer shall be applied. When a concrete is not cured, this has negative influence on final quality. For a shotcrete this is even more so, because of:
- thin layers that can dry out quickly, sometimes even to the substrate side
- the concrete is often accelerated, creating quick setting and high temperatures
- the ventilation air blowing by at constant and high speed

The concrete improver admixture shall be added to the concrete in the batching plant. It produces improved cement hydration through the whole cross section, better bond to the substrate and less risk of cracking. One reason for these advantages is that it is active from the very first second and a second reason is that subsequent layer application requires no removal of a spray-on membrane.

The admixture is compatible with the consistency control system. It also has the effect of improving concrete pumpability. The inter layer bond strength as checked by the contractor on a project in Saudi Arabia, showed 0.5 to 0.7 MPa without curing and above 2.0 with the concrete improver admixture. An increase of concrete density by 15% and compressive strength by 10% has also been recorded.

De Lustin Railway Tunnel in Belgium, used this system including the curing admixture. Some features and results can be listed as follows:
- shotcrete thickness 250 mm, with reinforcement
- mix design contained 420 kg cement with micro silica
- the consistency control system admixtures, including hydration control and concrete improver were used
- cores from the structure gave compressive strength 32 MPa after 28 days and 53 MPa after 90 day

5 Benefits of modern SFRS

The use of modern wet mix shotcrete produces benefits in different fields. The relative importance of such benefits may vary from one project to another. How to present advantages will depend on the basis of comparison. A comparison between traditional and modern methods is probably the best way of illustration. Dry mix shotcrete application is therefor used as a comparison basis, combined with mesh reinforcement. The benefits can then be identified within the following fields:
- improvements in quality and durability
- better safety and working environment
- improved technical properties
- time savings and improved economy

5.1 Quality and durability

For rock support, the Observational Method and one pass lining benefits will be partly lost, unless all applied shotcrete becomes part of the permanent support. Obviously, this calls for stricter requirements on quality assurance of the shotcrete. Modern SFRS is superior in this respect, for a number of reasons. To mention a few: Weight batching at a modern mixing plant, no mix design influence by the nozzle-man, pre-hydration can be completely excluded and target mix design is closely reproduced, due to low rebound (including steel fibres).

Currently, a number of projects are executed with required shotcrete compressive strength of 35 to 50 MPa. This is achieved by low w/c-ratio (normally, 0.4 to 0.45) and the use of modern accelerator- and admixture systems. Required toughness for SFRS may be as high as $I_{30} = 22$ according to ASTM 1018, or 700 Joule according to the French Norm SNCF. Silica fume is influencing a number of shotcrete properties favourably. One of the advantages is that the sulphate attack resistance capability is improved. The alternative use of a sulphate resistant cement type is actually giving poorer results [1].

Even more important is the resistance against reinforcement corrosion. It is a well known fact that mesh reinforcement has a high risk of poor compaction locally, behind the bars. Particularly, where mesh mats are overlapping (sometimes 3 to 4 layers!). When corrosion starts, the concrete cover is spalling and a rapid strength deterioration occurs. Research shows that if chlorides are available, it takes only 2% chlorides drawn on the cement weight, to initiate corrosion. When this is compared to steel fibre structures, 6% chlorides had not yet initiated corrosion [2]. In Norwegian sub-sea tunnels, there are records showing fully chloride saturated locations, without steel fibre corrosion [3].

Modern low pulsation shotcrete pumps with digitally controlled accelerator dosage systems and a hydraulic applicator, allow for a high and consistent concrete quality. Difficult access and operator exhaustion is not to the same extent as before, negatively influencing quality.

The influence of curing treatment (or not) on shotcrete quality and durability is often underestimated. Because of thin layers, surface roughness, ventilation air and sometimes high temperatures, curing must be ensured from the first minutes after application. The only practical approach so far, has been the spray-on membrane. Presently, an admixture is available, that covers the effect of the spray-on membrane with additional advantages. Improved hydration, better bond, improved compressive strength and reduced shrinkage, has been recorded in project applications and in tests.

5.2 Safety and Working Environment

The critical phase when tunnelling in poor ground, with a short stand-up time, is the first period after blasting a round, or by mechanical excavation when exposing fresh tunnel roof. Regardless of shotcreting method, installation of mesh is a highly manual work. This means that people have to venture below a tunnel roof, not yet fully supported.

With SFRS this is quite different. If need be, the first application can be quickly executed by robot over the muck pile, with the reinforcement included. The method also invites a full thickness shotcrete layer as a first step and any required rock bolts to be drilled through as a second stage. The advantage is that the bolt bearing plate is placed on the shotcrete surface, producing maximum load transfer between the two support elements. Additionally, the bolt installation takes place under a properly supported area.

Dust emission is partly a problem of shotcrete application. Depending on ventilation conditions and operator distance from the nozzle, the hygienic limits for total dust and for breathable quarts may be violated. Using the wet mix process, this is normally not a problem. The dry mix process produces substantially more dust in the nozzle area, but very often, as much at the location of the spraying machine. The sum of this dust emission may even cause problems further out in the tunnel [4].

5.3 Technical Properties

Especially in the period 1970 to 1985, a number of full scale model tests were made in Scandinavia and North America, comparing mesh- and steel fibre reinforcement [5] and [6]. The tests consistently showed, that steel fibres are giving equal, or better flexural strength and ductility (toughness). See Fig. 1 and 2.

In a tunnelling situation, where the mesh location cannot be controlled, the fact that SFRS has reinforcement in all sections, turns into a major technical advantage.

The fibres also have the effect of arresting part of the shrinkage forces, probably being the main reason why the bond to the substrate is normally improved.

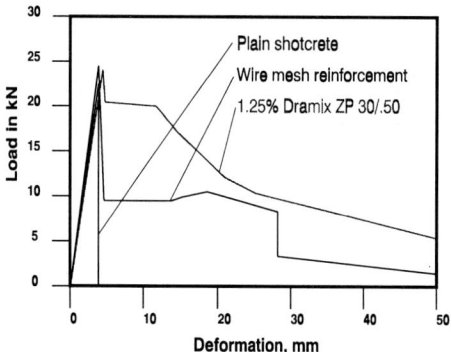

Figure 1

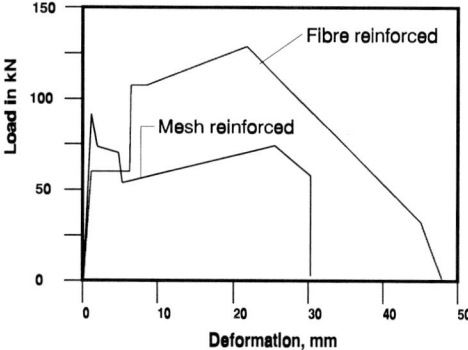

Figure 2

The shadow effect of mesh reinforcement has been mentioned. One way of reducing the problem, is to apply half of the planned thickness, then the mesh and finally the remaining shotcrete layer. This procedure should be a requirement in all cases involving mesh. When work sequencing allows, the planned layer thickness of SFRS may be applied in one operation. The potential problem of inter-layer bond and a laminated structure, may therefore be avoided. It is also much easier to accurately produce the thickness of shotcrete actually required.

5.4 Time and Economy

There are four main reasons why SFRS is saving time and thereby being a more economic support alternative in most cases:

1. Modern wet mix pumps and robot systems have a maximum output of about 20 m^3/h. This is especially useful, when available surface allows such high output.
2. **No time** spent on mesh installation. Additionally, there is no time lost in change of operations three times, as when using mesh. On blasted rock surfaces, theoretical thicknesses of 100 to 150 mm with mesh, normally requires about 50% extra concrete to cover the mesh.
3. Regardless of shotcrete method, application on mesh is increasing the rebound. Comparing the extremes, dry mix and mesh with wet mix and fibres, the difference in rebound rate can easily amount to twenty percentage points (reduction of rebound rate by 66%).
4. Provided that the high quality SFRS is taken advantage of in the design approach, direct production of permanent support saves project time and cost.

An illustration of the above are made by calculations of dry mix application using mesh reinforcement and wet mix SFRS. The figures demonstrate the importance of looking into project cost figures, rather than focusing upon part cost like materials and equipment investment. The printouts are enclosed as appendix A (four pages).

The basis is a 60 m^2 tunnel, 2000 m long, drill and blast excavation, 4 m round length, 100 mm theoretical shotcrete thickness applied per round, European cost level and Swiss Francs calculation currency. Dry mix rebound 30% and wet mix rebound 10%. The summary figures are illustrated in Fig. 3 (complete data in appendix. A).

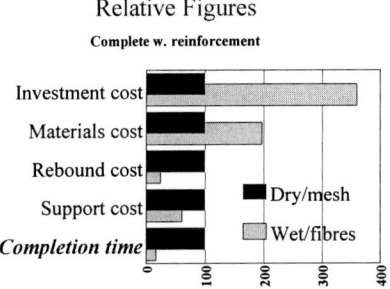

Figure 3

Investment cost (in Fig. 3) is the value at the start of construction of the complete equipment package. The dry mix equipment value used is CHF 125'000, while the wet mix equipment is 450'000. Depreciation of 1.5% per month and interest rate of 8% pro annum are the same in both alternatives. *Materials cost* means direct cost of all ingredients of the mix design, including accelerator added in the nozzle. The actual figures are CHF 166 and 327 per m^3 for dry and wet respectively. Wet mix without fibres would be CHF 255 per m^3. *Rebound cost* contains the loss of materials, cost of extra spraying to compensate and the cost of rebound removal. The difference in rebound cost (CHF 800'000), could pay the wet mix equipment investment 1.8 times. Even if the site delay time value is put at zero, the rebound cost difference is still CHF 480'000, or equal to 70% of the dry mix materials cost. *Support cost* means all direct cost in the shotcrete- and reinforcement application, **not** including indirect cost-sharing of infrastructure, administration, other equipment used or delayed, other workers involved etc. If a cost allocation of a modest CHF 400 per shift hour is applied for this, the total cost of SFRS drops to **42% of dry mix and mesh**. This saving more than covers the total cost of SFRS application through the whole tunnel. The main reason for such dramatic figures, is shown in the last bars, *Completion time*: The SFRS method in this case only needs **17%** of the time, to finalise the same support volume.

The above illustrates the savings potential on a contractor's side. For a client, the potential savings are often several times the above level. It is therefore, primarily, the client side that will suffer the consequences of not optimal solutions.

6 Conclusion

The set of arguments in favour of modern high performance SFRS is today very strong. In all fields of evaluation, like quality, safety, environment and time and economy, the conclusions are consistently the same.

It is sometimes argued that the application of such "high cost" modern methods is too expensive and may turn out too complicated. In reality, this technology is easier to control in terms of quality assurance. The part cost focus should be avoided, since the wrong method will generate total cost much higher than the direct part cost savings.

The key to a successful use of modern rock support technology lies with the client and his consultants. The whole strategy from rock support design, through terms of contract and contractor incitement for shorter project time, has to be designed for efficiency. The major part of the potential economical benefits will remain on the clients hand, also if the contractor is given a share.

7 References

1. Scherer, J. and Fidjestöl, P (1993) *Microsilica Betone unter Sulfatangriff*, Report (intermediate after 3 years testing) by Elkem a/s Chemicals and Meynadier AG
2. Mangat P.S., Molly B.T. and Gurusamy K., *Marine Durability of Steel Fibre Reinforced Concrete of High Water/Cement Ratio*, full reference not available.
3. NOTEBY Consulting Engineers (1986) *Investigation of the shotcrete used in the Kaarstoe Tunnels*, Report No. 2, Project No. 21303
4. Ono, K. (1995) *Health and Safety in Shotcreting*, preliminary report to the International Tunnelling Association Working Group on Shotcrete, Stuttgart
5. Morgan D. R. and Mowat D. N (1982) *A Comparative Evaluation of Plain, Mesh and Steel Fibre Reinforced Shotcrete*, Hardy Associates, International Symposium on Fibre Reinforced Concrete - Detroit
6. Holmgren J (1983) *Tunnel Linings of Steel Fibre Reinforced Shotcrete*, 5th International Congress on Rock Mechanics, Melbourne

INPUT

K. Garshol
Jan. 1995

BASIC PROJECT INFORMATION		User Input			
Calculation number:		1			
Calculation Date:		15.05.96			
Project name:		Appendix A	Dry mix shotcrete with mesh		
Project location:					
Contractor:					
Client:					
Shotcrete method calculated:					
Wet-mix (Enter 1):	U N A V A I L A B L E				
Dry-mix (Enter 2):		2		DRYMIX CALCULATED !	
Average rebound rate, %:		30			
Steel fibres (Enter 4):					
Mesh reinforcement (Enter 5):		5		MESH REINFORCEMENT !	
Project duration:					
Start date (dd.mm.yy):		01.01.95			
Planned end date (dd.mm.yy):		01.06.96			
Duration, months:		17			
Excavation method:					
Drill & Blast (yes/no):		Yes			
Mechanical (yes/no):		No			
Compressed air needed, m3/min.:		10			
Electricity needed, kW:		20			
Expected total shotcrete volume, m3:		4'320			
Expected total shotcrete area, m2:		43'200	=> Average t, mm:	100	
Calculation currency:		CHF	Realistic shotcrete thickness ?		
MATERIALS					
Concrete mix design, kg per m3			User input		
(in batching plant)			unit prices		
Cement, kg:		400	0.150	Type:	OPC
Microsilica, kg					
Sand/aggregate I, kg		1700	0.040	Fraction:	0-16 mm
Sand/aggregate II, kg					
Sand/aggregate III, kg					
Sand/aggregate IV, kg					
Admixture I, kg					
Admixture II, kg					
Admixture III, kg					
Admixture IV, kg					
Meyco TCC 735, kg				Type:	
Added in the nozzle, kg per m3:					
Accelerator I, kg		32.0	1.20	Type:	Powder aluminate
Accelerator II, kg					
Other materials:					
Curing membrane, kg/m2:		0.5	4.07	Type:	Spray-on membrane
Mesh reinforcement, kg/m2		6.0	1.50	Type:	
EQUIPMENT					
01.01.95	Start value of compl.				
equipment package:	CHF	125'000			
Consisting of:					

Depreciation, % per month:	1.5					
Interest rate, % pro annum:	8.0				NB, Input !	
Cost of spare parts per m3:	3.00	Cost of maintenance per m3:			5.00	
Other costs / m3:	14.00	(Electricity, compressed air, batching plant)				
SHOTCRETE CREW/TRANSPORT						
Average manhour cost:	70.00					
Number of men per crew:	3					
Shifthour cost per concrete transport unit:	150.00					
Transport units/shotcrete crew:	1					
SITE DELAY TIME						
Cost/shift shotcrete- or mesh application, NOT covered above:	400.00					
PRODUCTION CAPACITIES						
Hours per shift:	8.0					
Shifts per day:	3.0					
Days per week:	5.0					
Average m3 through equipment, per shifthour:	2.00					
Mesh, m2/shifthour:	11.00					
OTHER ITEMS						
Rebound removal, cost/m3:	2.00					

OUTPUT

UNIT COSTS SHOTCRETE					
Materials					
m3 cost of shotcrete materials including accelerator:				166	
Equipment					
Depreciation of equipment package:			31875		
Interest on equipment value:			12360		
Equipment capital cost:			44235		
Capital cost per m3 of shotcrete:				10	
m3 cost of spare parts, maintenance and energy:				22	
Direct production cost					
m3 cost for shotcrete crew:			105		
m3 cost for concrete transport:			75		
m3 cost manpower and transport:				180	
TOTAL COST / m3 OF SHOTCRETE THROUGH EQUIPMENT				379	
TOTAL COST / m3 OF SHOTCRETE ON THE SURFACE				542	
UNIT COSTS IF MESH REINFORCEMENT AND CURING					
Materials					
m2 cost of mesh reinforcement:			9		
Direct production cost					
m2 cost of mesh installation:			19		
TOTAL COST / m2 INSTALLED MESH				28	
Cost of curing membrane	/m2			2	
TOTAL PROJECT FIGURES					
Total cost of shotcrete, rebound included				2'340'451	
Total cost of mesh reinforcement				1'213'527	
Total cost of curing				87'912	
Total time cost, not included in above figures				2'805'195	
Total project cost of (reinforced) shotcrete application				CHF	6'447'085
Total cost of rebound included with:				CHF	1'078'716
Total shifthours shotcrete (+ reinforcement)				Shift h	7'013
Extra time caused by rebound, included with:				Shift h	926
this would take in consecutive shifts:				Months	13.5

New admixtures for high performance shotcrete

INPUT							K. Garshol	
							Jan. 1995	
BASIC PROJECT INFORMATION			User Input					
Calculation number:			2					
Calculation Date:			15.05.96					
Project name:			Appendix A	Wet mix shotcrete with steel fibres				
Project location:			(Comparison to traditional dry-mix with mesh)					
Contractor:								
Client:								
Shotcrete method calculated:								
	Wet-mix (Enter 1):		1		**WETMIX CALCULATED !**			
	Dry-mix (Enter 2):	U N A V A I L A B L E						
	Average rebound rate, %:		10					
	Steel fibres (Enter 4):		4		**STEEL FIBRE REINFORCEMENT !**			
	Mesh reinforcement (Enter 5):							
Project duration:								
	Start date (dd.mm.yy):		01.01.95					
	Planned end date (dd.mm.yy):		01.01.96					
	Duration, months:		12					
Excavation method:								
	Drill & Blast (yes/no):		Yes					
	Mechanical (yes/no):		No					
Compressed air needed, m3/min.:			15					
Electricity needed, kW:			20					
Expected total shotcrete volume, m3:			4'320					
Expected total shotcrete area, m2:			43'200	=> Average t, mm:		100		
Calculation currency:			CHF					
MATERIALS								
Concrete mix design, kg per m3				User input				
(in batching plant)				unit prices				
	Cement, kg:		425	0.150	Type:	OPC		
	Microsilica, kg		25	0.500	Type:	Densified		
	Sand/aggregate I, kg		495	0.035	Fraction:	0-4 mm		
	Sand/aggregate II, kg		1155	0.040	Fraction:	4-8 mm		
	Sand/aggregate III, kg							
	Sand/aggregate IV, kg							
	Admixture I, kg		4.0	2.50	Type:	Superplasticiser		
	Admixture II, kg		3.6	2.00	Type:	Plasticiser		
	Admixture III, kg		2.7	6.58	Type:	Hydration control		
	Admixture IV, kg							
	Meyco TCC 735, kg		5.0	2.50	Type:	Internal Curing		
	Steel-fibres, kg		40.0	1.80	Fibre type:	Dramix 30/.50		
Added in the nozzle, kg per m3:								
	Accelerator I, kg		22.5	3.00	Type:	Alkali-free		
	Accelerator II, kg							
Other materials:								
	Curing membrane, kg/m2:							
EQUIPMENT								
	01.01.95	Start value of compl.						
	equipment package:		CHF	450'000				
	Consisting of:							

Depreciation, % per month:	1.5			
Interest rate, % pro annum:	8.0			NB, Input !
Cost of spare parts per m3:	3.00	Cost of maintenance per m3:		5.00
Other costs / m3:	15.00	(Electricity, compressed air, batching plant)		

SHOTCRETE CREW/TRANSPORT

Average manhour cost:	70.00
Number of men per crew:	1
Shifthour cost per concrete transport unit:	150.00
Transport units/shotcrete crew:	2

SITE DELAY TIME

Cost/shifht shotcrete- or mesh application, NOT covered above:	400.00

PRODUCTION CAPACITIES

Hours per shift:	8.0
Shifts per day:	3.0
Days per week:	5.0
Average m3 through equipment, per shifthour:	4.13

OTHER ITEMS

Rebound removal, cost/m3:	2.00

OUTPUT

UNIT COSTS SHOTCRETE

Materials			
m3 cost of shotcrete materials including accelerator:		327	
Equipment			
Depreciation of equipment package:	81000		
Interest on equipment value:	32760		
Equipment capital cost:	113760		
Capital cost per m3 of shotcrete:		26	
m3 cost of spare parts, maintenance and energy:		23	
Direct production cost			
m3 cost for shotcrete crew:	17		
m3 cost for concrete transport:	73		
m3 cost manpower and transport:		90	
TOTAL COST / m3 OF SHOTCRETE THROUGH EQUIPMENT		**466**	
TOTAL COST / m3 OF SHOTCRETE ON THE SURFACE		**518**	

UNIT COSTS IF MESH REINFORCEMENT AND CURING

Materials		
Direct production cost		
Cost of curing membrane	/m2	

TOTAL PROJECT FIGURES

Total cost of shotcrete, rebound included	54'000	2'236'141	
Total cost of curing (Internal curing included with:)			
Total time cost, not included in above figures		464'891	
Total project cost of (reinforced) shotcrete application		CHF	**2'701'032**
Total cost of rebound included with:		CHF	271'927
Total shifthours shotcrete (+ reinforcement)		Shift h	**1'162**
Extra time caused by rebound, included with:		Shift h	116
this would take in consecutive shifts:		Months	**2.2**

4 ADVANCED EXPERIENCES WITH HIGH PERFORMANCE ALKALIFREE NON-TOXIC POWDER ACCELERATOR FOR ALL SHOTCRETE SYSTEMS

D. Mai
Sika AG, Zurich, Switzerland

Abstract
Modern tunnel construction methods have gained increased importance on a world wide basis. Demand for both, the wet and the dry shotcreting methods, is steady on the increase. Currently, strong caustic alkaline accelerators are being used with a resultant significant reduction in concrete strength at later ages.
Aside from the demand for higher quality and improved economy, today, there must also consideration be given toward ecological viewpoints as well as toward a safer working environment. The described new alkalifree powder accelerator, applicable in all wet and dry shotcrete processes, can be viewed to meet these demands and ensures an improved ecology in tunnel construction.
The new accelerator is not caustic, not hazardeous and not toxic (according to European regulations). The transport declaration is "No dangerous goods". Tests and experiences gathered in a research tunnel and on several construction sites have shown positiv results and will be reported. Compared to conventional accelerators, no ultimate strength reductions can be observed. On the contrary, increased ultimate strength values can be possible.
Leaching tests with de-ionized water have indicated significantly less alkalies and calcium hydroxide in the water that was in contact with shotcrete samples accelerated by the powder. That means reduced clogging of tunnel drains and conduits, less alkali pollution of ground waters and less contamination of shotcrete waistes during long term depositing.
The simultaneous development of a practical dosage equipment ensures a simple handling, also the continuous precise addition of powder accelerator is guaranteed.
Keywords: alkalifree, dry spraying, ecology, non-toxic, powder accelerator, shotcrete, strength increase, wet spraying.

1 Introduction

Shotcrete is an essential part of modern underground construction and becomes increasingly important in the future, both, to stabilize the rock and to establish permanent shotcrete linings. To meet these structural requirements high performance shotcrete is essential. In addition to the high quality standards, other advantages as improved working conditions for the workers in the tunnel and reduced environmental pollution, are gaining increased importance on a world wide basis. Shotcrete accelerators play an important role to guarantee these objectives.

2 Shotcrete accelerators

2.1 General overview

Ever since shotcrete processes have evolved, different accelerators were and are still in use or under development, depending on the shotcrete process itself and the national regulations and habits. All these accelerators, listed in Fig. 1, differ not only in their chemical characteristics, their technical performance and in economic aspects, but also significantly in causticity, toxicity, transport and storage, and other ecological viewpoints as working environment and emission of critical substances. Some even contain substances that are known as detrimental to health.

Today, the most common used shotcrete accelerators, liquid and powder, are predominantly composed of alkali aluminates and in some countries of alkali silicates. They both contain strong aggressive and caustic solutions and the dis-

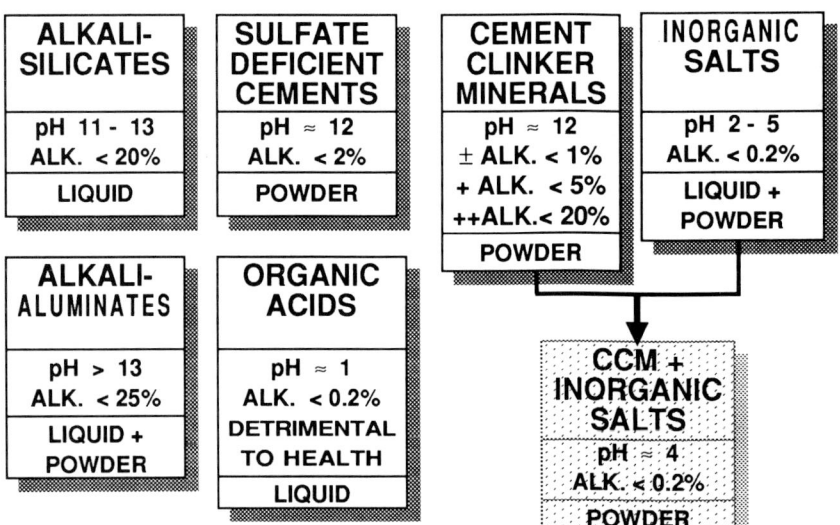

Fig. 1: Set accelerators and cements for shotcrete application.

advantages of these effective but also hazardeous products are well known:

- High pH values and thus a potential hazard
- High loss of strength when compared to not accelerated concrete mix
- High final shotcrete quality can often only be obtained by silica fume addition

2.2 Alkalifree non-toxic powder accelerators

Alkalifree non-toxic powder accelerators, based on clinker minerals and inorganic salts, are non-caustic and have pH values between 4 and 6. That leads to improved health and safety at work. Accidents and serious injuries can be prevented and potential dangers during transport and storage can be avoided. Alkalifree powder accelerators do not increase the alkali content of the sprayed concrete. That results in higher strength without any loss of ultimate strength. Reduced lining thickness helps decreasing the consumption of valuable aggregate/sand raw material resources. Less rebound, and rebound not polluted by alkaline set accelerators, ensures not only better economy but also improved ecology in shotcreting. Alkalifree powder accelerators compact the microstructure of the cement matrix of the shotcrete and, therefore, decrease the water permeability and reduce the quantity of elutable calcium hydroxide compared to other set accelerators. This, e.g., reduces the clogging of drainage pipes and the discharge of highly alkaline waste waters into the drainage system, and also into ground water and river systems.

3 Technical specifications of alkalifree non-toxic powder accelerators

3.1 Chemical reactivity

It has to be assumed that an alkalifree powder accelerator based on several cement clinker minerals and inorganic salts, reacts similar to the majority of all shotcrete accelerators. The rapid hardening of the shotcrete seems to be mainly a result of the significantly influenced course of reaction of the cement clinker phase tricalcium aluminate (C3A). However, cement hydration is a very complex subject based on scientific knowledge published in the past 80 years. When adding accelerators to the cement, the observation of a increased rate of reactions can be compared as looking into an aquarium filled with fish that feed on each other.

I don't want to go into details but I'd like to present some of the auhor's insitu measurements on hydration temperatures and setting times of sprayed concrete and laboratory mixes, accelerated with the new alkalifree powder accelerator.

In Fig. 2 the very strong increase of the heat of reaction of the alkalifree powder during the first eight hours of hydration, compared to the conventional accelerators, can be observed. This could be an indication for the good early strength development under low temperature spraying conditions.

Fig. 3 shows initial and final set results of penetrometer lab tests. Also Portland cement HS (low C3A≈1%) can be accelerated with the alkalifree powder accelerator as fast as other Portland cements, or even faster. This clearly indicates an advantage compared to conventional shotcrete accelerators. In the European

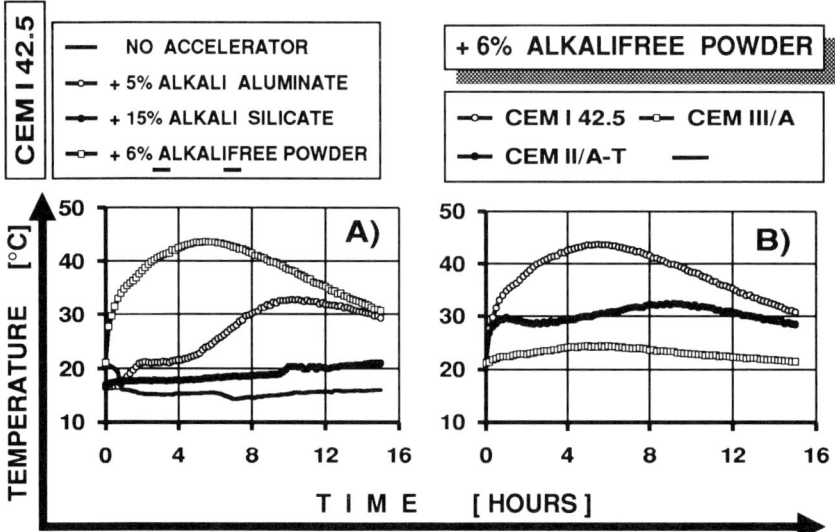

Fig. 2: In situ measurements on hydration temperatures. Dry spraying tests in test gallery. Mix type: PC 480 kg/m^3; sand 0/3 mm predried.
A) Influences of accelerator type.   B) Influences of cement type.

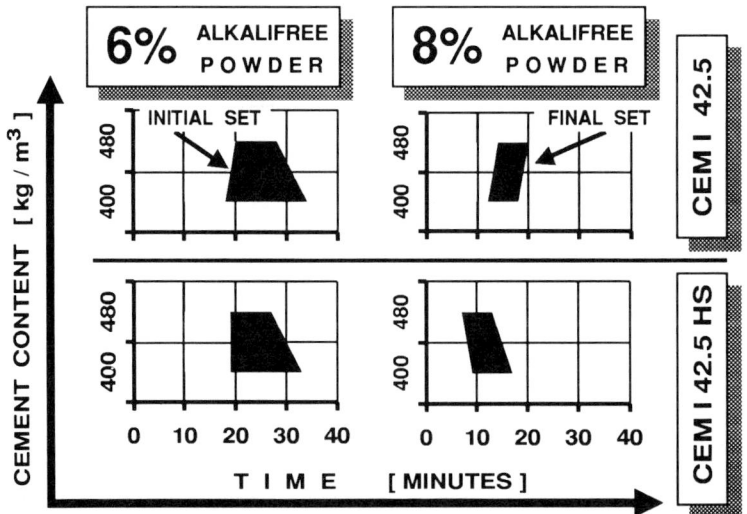

Fig. 3: Penetrometer laboratory tests. Standard mix: cement + sand 0/2 mm (W/C=0.55); 1% stabilized super plasticizer.

Alpine region many tunnels will be penetrated by mountain waters that have high sulphate contents. Therefore, preferably low C_3A-cements will be used more often.

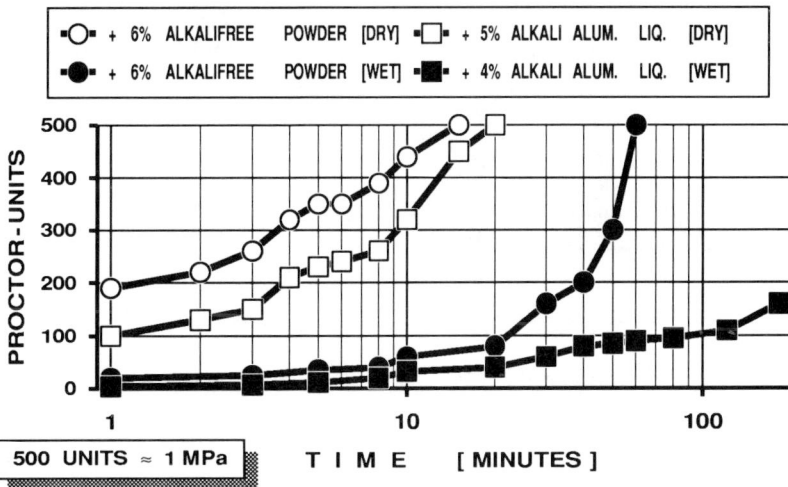

Fig. 4: Insitu early strength measurements (0 to 2 hours), detected by a portable needle penetrometer tester (Proctor Tester).
Dry mix : CEM I 42.5;   480 kg/m^3; sand 0/3 mm predried
Wet mix : CEM I 42.5;   450 kg/m^3; aggregates 0/8 mm; W / C = 0.50.

3.2 Early strength and ultimate strength developments

Results of early strength development, detected with a portable needle penetrometer are shown in Fig. 4. Under equal conditions (mix; comparable accelerator dosage; spraying equipment; temperatures; nozzle man) the alkalifree non-toxic powder accelerator acts significantly faster than the compared alkali aluminates. This can be observed in both, the dry and the wet spraying process.

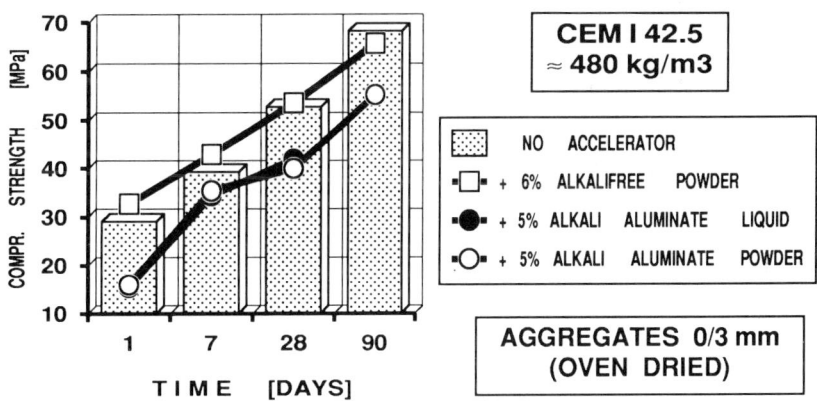

Fig. 5: Ultimate strength developments observed in the dry spraying process.

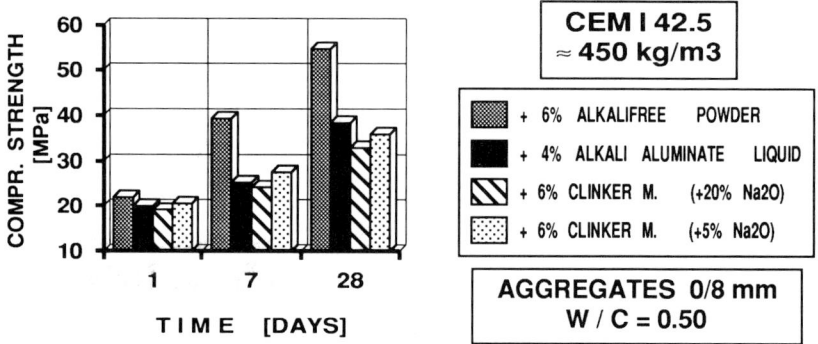

Fig. 6 : Ultimate strength developments observed in the wet dense process
(Total alkali contents is expressed in Na2O wt.-%).

Ultimate strength results are represented in Fig. 5 and Fig. 6. The results of dry spraying (Fig. 5) confirm that no ultimate strength reduction, compared to conventional alkali aluminate accelerators, can be observed. The results of the wet spraying process (Fig. 6), compared to competitive powder accelerators based on clinker minerals containing larger amounts of alkalies, clearly demonstrate the advantage in using an alkalifree powder accelerator.

3.3 Rebound and dust emission

Fig. 7 gives an example for rebound reduction and dust emission decrease in the dry spraying process. All other spraying parameters, except the accelerator type, had been kept constant. Therefore, these results show directly the positive

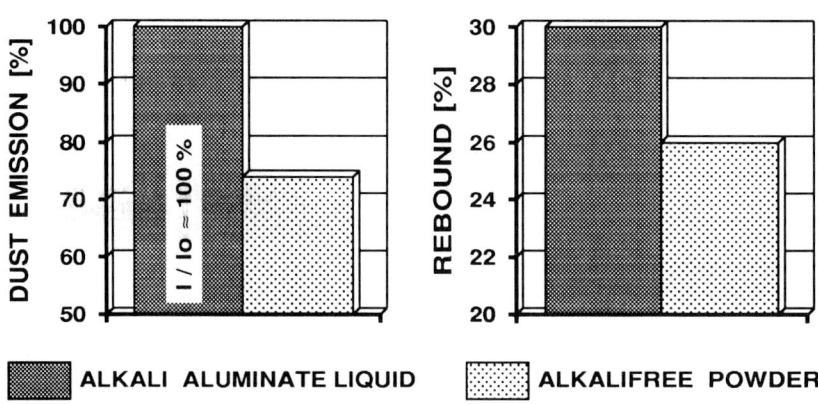

Fig. 7 : Rebound and dust emission measurements in the dry spraying process.
Dry mix : CEM II / A-V ; 370 kg / m^3; aggregates 0/16 mm; mix humidity ≈ 3%; temperature < 10 °C.

influences on rebound and dust emission of the alkalifree non-toxic powder accelerator compared to the conventional types.

3.4 Leaching tests and effluent values

All values presented here have been determined from measurements in a leaching test cell, where shotcrete discs have been leached out by de-ionized water with a comparably low flow rate of 300 ml /h. For chemical analyses water samples have been taken after 2 hours to 3 days leaching time.

Fig. 8 shows cumulative effluent values representing the wet process, sprayed with alkalifree powder, alkali containing powder and alkali aluminate liquid accelerators. The results can be described as following :

- With an alkalifree non-toxic powder accelerator no additional soluble alkalies are introduced into the shotcrete mix
- Also the microstructure of the cement matrix of the hardened shotcrete is more compacted and therefore, the quantity of elutable calcium hydroxide is reduced
- That means, an alkalifree non-toxic powder accelerator significantly reduces the clogging of drainage pipes and the discharge of highly alkaline waste water into drainage systems

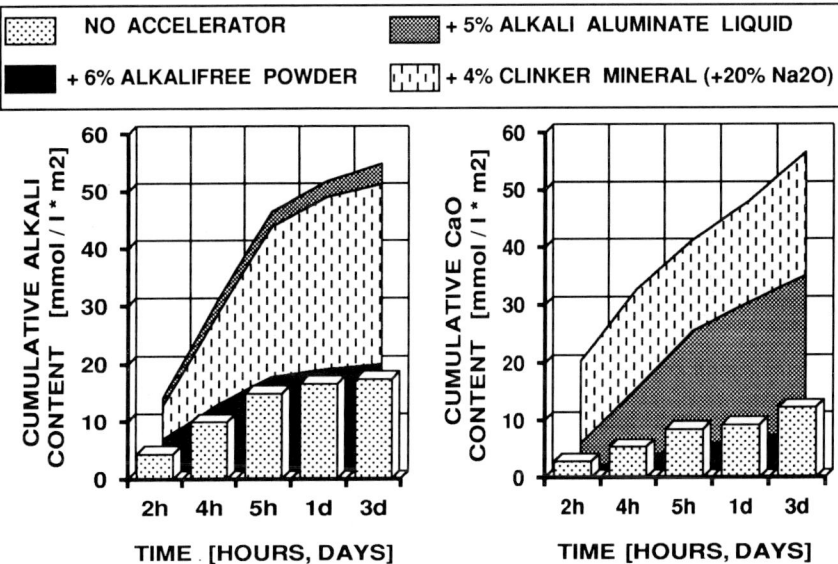

Fig. 8 : Elutability of alkalies and calcium hydroxide of concrete, sprayed in the wet process, determined in a leaching test cell developed by the author.
Wet mix : CEM I 42.5; 450 kg/m^3; aggregates 0/8 mm.
Leaching cond.: flow rate 300 ml/h; total shotcrete surface 1800 cm^2.
Total alkalies (Na$_2$O + K$_2$O) expressed in Na$_2$O equivalents.
Total calcium hydroxide (Ca(OH)$_2$) expressed in CaO equivalents.

To understand these clogging phenomena, the following two mechanisms, in simplified description, are responsible for the lime deposits in drainage pipes :

Besides the hydrated clinker phases a variety of hydroxides of varying solubility exists in the hardened cement matrix. The largest amount, in quantitative terms, is the calcium hydroxide content. Depending on the density of the matrix, this is washed out by ground water and precipitated when in contact with CO_2 from the air or water to form limestone that clogs drainage pipes. If the matrix is permeable and porous, other soluble alkalies from conventional accelerators, such as sodium and potassium hydroxide may also be leached out by the water flow. The solubility of calcium hydroxide then decreases and lime is deposited in the drainage system at even earlier stage.

4 Tunnel site results

As an example Fig. 9 contains ultimate strength results of the Piora Pilot Tunnel in Switzerland. This tunnel site belongs to the **N**ew **A**lp **T**ransversal project (NEAT). Two different cements have been accelerated with alkalifree non-toxic powder accelerator and conventional alkali aluminate liquid accelerator. The positive influence of the alkalifree powder produces a significant reduction in the strength gap between core samples and reference cubes for both tested cement types.

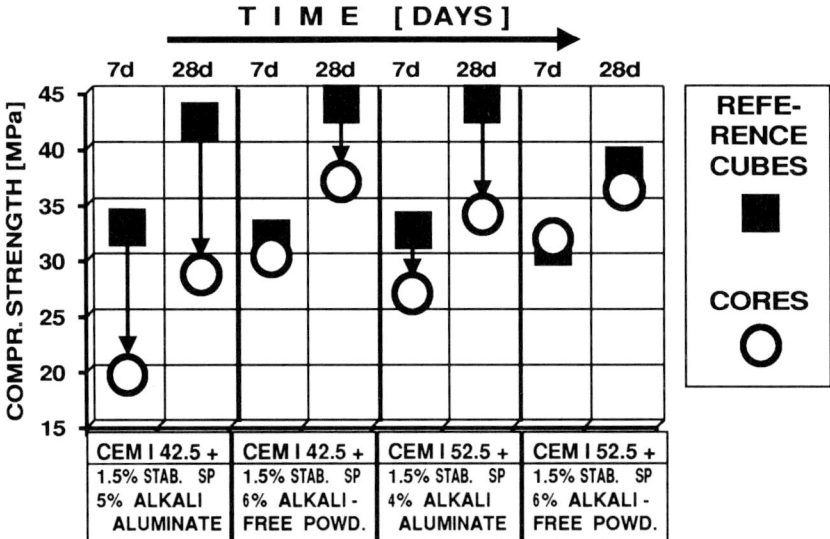

Fig. 9 : Ultimate strength development at site "Piora Pilot Tunnel", Switzerland.
<u>Wet mix :</u> PC 450 kg / m^3; aggregates 0/8 mm; 20 kg / m^3 steel fibres; 1.5% stabilized super plasticizer.

PART TWO
APPLICATIONS

5 SPRAYED CONCRETE IN CONJUNCTION WITH ELECTRO CHEMICAL PROTECTION OF CONCRETE

K. Dykes
Makers Industrial Ltd, Warrington, UK

Abstract
Sprayed concrete has been used for many years to repair and protect numerous concrete structures, encompassing all types of members. With the introduction of electro chemical, i.e. cathodic protection, and the desalination/realkalisation processes to the repair sector of the construction industry, sprayed concrete has been adapted to be used in conjunction with these systems. This paper is dedicated to the repair and protection of reinforced concrete structures.
Keywords: corrosion, cathodic protection, desalination, realkalisation, repair, testing

1 The Problem

Corrosion is probably the most serious cause of deterioration of reinforced concrete structures. Buildings and bridges are suffering from corrosion due to the ingress of carbon dioxides, deicing salts and salts cast into the structures, i.e. calcium chloride as an acceleration agent during its construction.

A Department of Transport report showed that over half a billion pounds worth of electro chemical protection will be required before the next millennium to preserve the motorway and trunk road system in England and Wales. Comparable problems exist on other reinforced concrete structures all over the world.

The reinforcement steel is usually protected from corrosion when it is embedded in concrete as the alkalinity of the concrete stops the acidic corrosion process. The protection to the steel can be broken down by carbon dioxide gas, which is acidic, or by salt. When corrosion of the steel occurs it forms corrosion products (rust) which expands and cracks the concrete, inducing the concrete to spall/delaminate and structural weakening. The corrosion process is electro chemical.

2 Cathodic Protection

The process of cathodic protection goes back to Sir Humphrey Davy's discovery of the process in 1824. It has been used in marine applications for over a century and was first applied to bridges when the corrosion process was discovered in the early 1970's. There is now over twenty five years of experience of its application to various types of structures.

Cathodic protection works by producing electrons from an external power source (a D.C. power supply) via an external anode installed on or in the system. When the electrons flow between the anode and the reinforcement steel it makes the corroding structure into a cathode, where the corrosion cannot occur. Hence the term Cathodic Protection (Fig. 1).

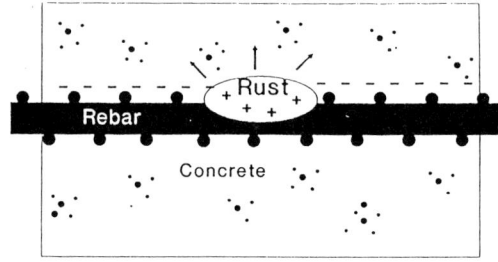

Corrosion proceeds by forming + and - areas on steel

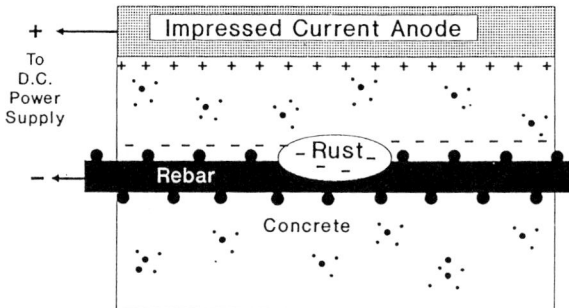

Cathodic protection makes all the steel "-"

Fig. 1 Cathodic Protection

There are a number of cathodic protection systems available. These include conductive coatings applied to the surface of the concrete, metal probes embedded into the concrete and expanded mesh embedded in a sprayed concrete overlay onto the surface of the concrete.

3 Realkalisation

Realkalisation is a technique to introduce an alkali solution into the cores concrete to the reinforcement to arrest and prevent further deterioration due to the carbonation process.

Realkalisation is an electrochemical method to restore alkalinity that has been lost during the carbonation process. An anode mesh is suspended above the surface of a concrete substrate. Tests take place to ensure that there is continuity within the reinforcing bars inside the concrete and connections made at regular intervals (Fig. 2).

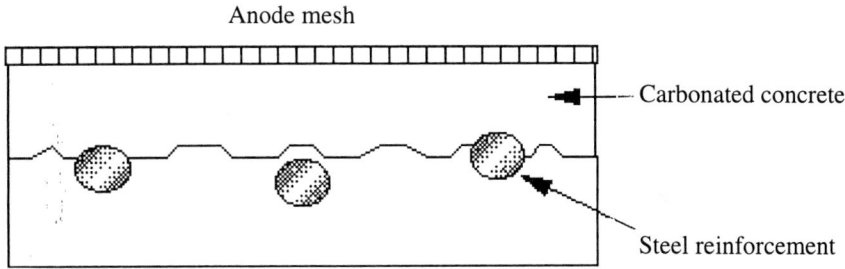

Fig. 2 Realkalisation

A fibre glass tank is fixed to the concrete encapsulating the anode mesh and filled with a sodium carbonate or calcium hydroxide solution at a calculated molar solution. A current density of 1 amp per m^2 is applied to the anode. Sodium or calcium disassociate and are attracted towards the cathode. In transport they carry their respective $Na_1 CO_3$ or $CA(OH)_2$ particles through the pour water towards the reinforcement. (Fig. 3).

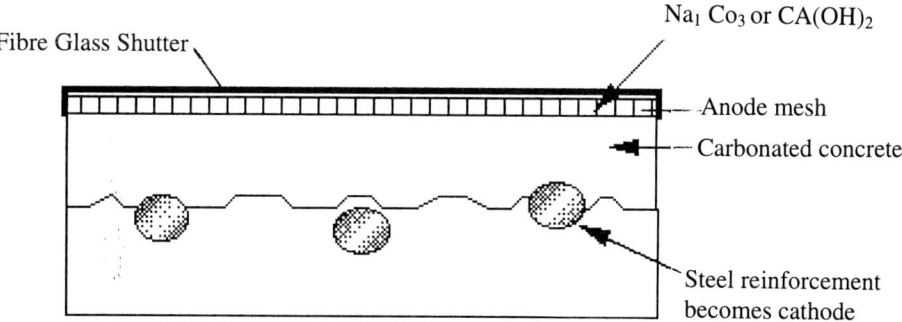

Fig. 3 Realkalisation

Soon after the system is powered up there is a drop in resistance between the anode and cathode. The disassociated metal ions are attracted towards the cathode. It takes between 48 and 72 hours for them to reach the bar. Simultaneously hydroxyl ions are formed at the cathode which further increases the alkalinity in the vicinity of the reinforcing bar. (Fig. 4).

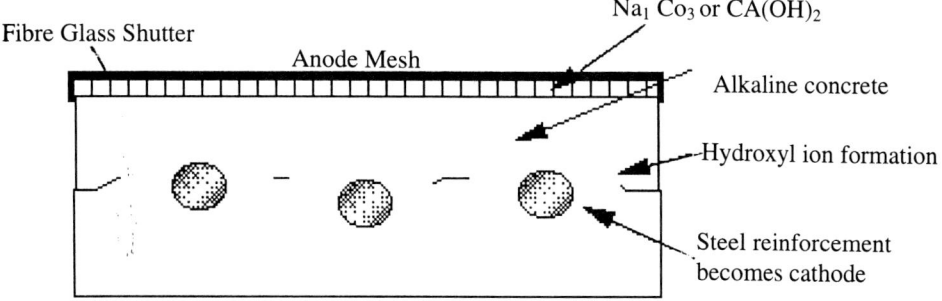

Fig. 4 Realkalisation

4 Desalination

Desalination is a technique to remove ingressed or cast-in chlorides in order to arrest deterioration in affected structures.

The passive oxide layer around reinforcing bar can be attacked by two forms of chloride contamination:
- Cast in chlorides used as an accelerator in the construction stage.
- Chlorides ingressed from an external source such as road salt.

Chlorides locally break down the protective Fe_1O_3 layer in isolated anodic sites. (Chlorides, when they come into contact with reinforcing steel in sufficient quantities can lead to localised breakdown of the alkaline passive film around the bar, producing small pockets of intensive corrosion. These are called anodic sites or pitting corrosion.) Macro cell corrosion ensures.

This corrosion can be best understood if we think of a battery with a constant short circuit:
- Steel dissolves
- Free electrons pass to cathode
- Oxygen + water + electrons product hydroxyl ions
- Hydroxyl ion transport charge back to the anode
- Rust products are formed

An anode mesh is applied to the substrate encapsulated in a fibre glass tank. The tank is filled with water which acts as an electrolyte. A current density of 1 amp per m^2 is applied to the anode. The resistance between the anode and the cathode drops. The chloride ions are attracted towards to positively charged anode. They pass into the water which is changed on a regular basis. Hydroxyl ions are formed at the bar in the same manner as realkalisation. (Fig. 5)

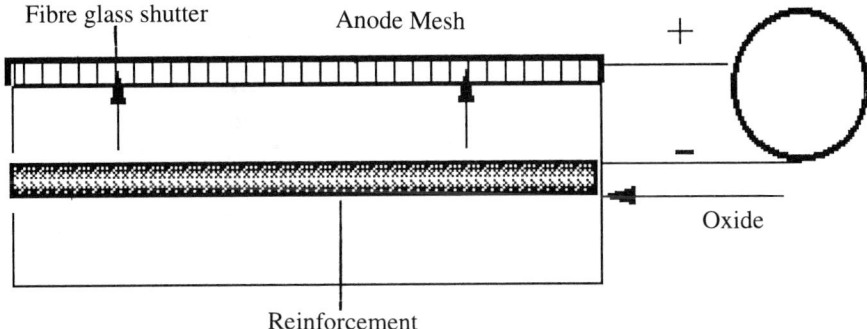

Fig. 5 Desalination

5 Testing

Prior to embarking on any repair project, an in-depth survey must be undertaken to determine the extent of contamination present in the structure.

The testing will generally consist of some of the following:-
- Graduated dust samples to ascertain chloride content, chloride gradients and cement content
- Petrographic analysis of core samples to identify aggregates used in construction, depth of carbonation, presence, if any, of alkali silica reaction
- Carbonation depth tests on site to confirm petrographic analysis
- Sodium and potassium content of concrete in 25 mm increments to a depth of 75 mm
- Continuity tests for steel reinforcement

For the Realkalisation and Desalination processes, testing is ongoing during and upon completion to provide the effectiveness of the system.

6 The Repair Process

On completion of the testing and results, all identified loose and delaminated areas of concrete are removed, preferably by hydro demolition techniques, to avoid any fracturing of aggregates in adjacent areas.

The exposed steel reinforcement is surveyed and if sectional loss is considered to be significant, the affected areas are replaced. The reinstatement of the areas is undertaken using sprayed concrete (wet or dry process), performed by a certified Nozzleman, ensuring full compaction. (Fig. 6) The sprayed concrete used for the patch repairs must be a reasonably matched electrical (ionic) conductivity material, with a similar density to concrete.

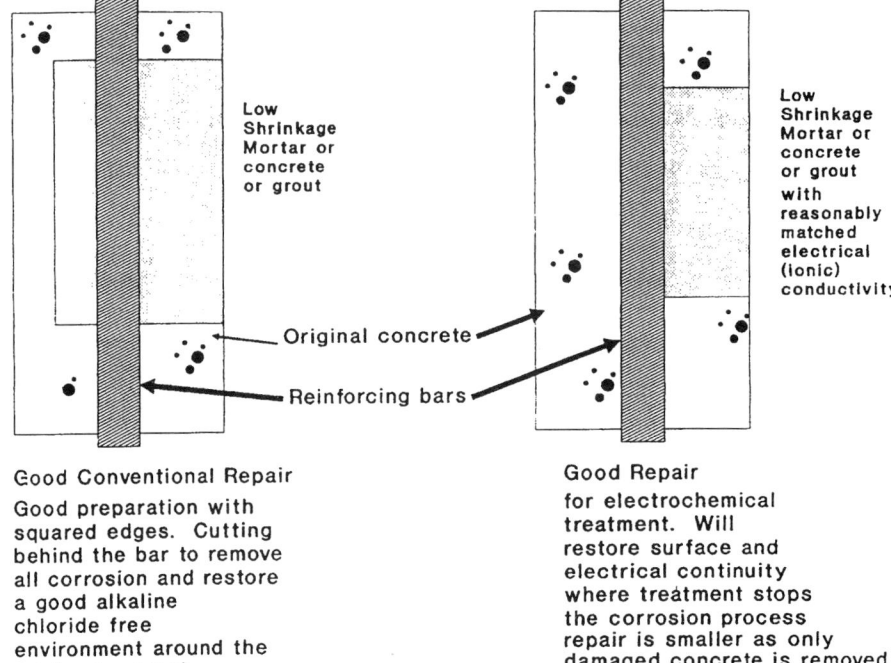

Fig. 6

This is particularly important in the realkalisation and desalination processes to allow the ingress of the alkali fluid to reach the reinforcement steel, or the egress of chloride ions to the external anode respectively. The worlds leading construction material manufacturers recognised the requirement for a sprayed concrete material compatible with the relevant chemical processes and have all formulated systems to suit.

With regard to the realkalisation and desalination processes, the internal connection to the steel reinforcement is made and the external shutters, incorporating the anode, are installed. The system is then connection to the computer monitoring system and the process is undertaken for the required period: realkalisation - four days, desalination - four to six weeks.

Upon completion of the process the shutters are removed and further testing is undertaken to provide the effectiveness of the process. The complete structure is then either painted with a protective coating or a 25mm thick sprayed concrete (wet or dry) layer is applied with either an as-shot finish or a steel trowel finish, dependent upon the clients requirement.

The cathodic protection process requires the internal connection to the steel reinforcement and the fixing of an external titanium mesh to the structure all connected to a permanent monitoring system that remains active for the life of the structure. The monitoring of the system can be undertaken remotely via a modern system to an office computer (Fig. 7).

Once the system is installed and checked, the external anode mesh is then protected by a sprayed concrete overlay, circa 25mm, and finished to the clients requirements.

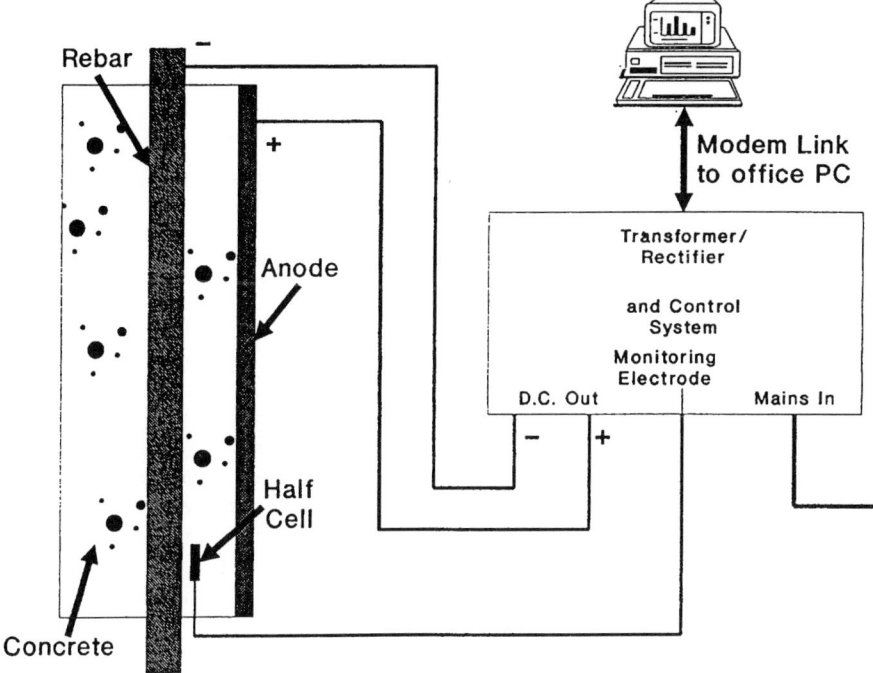

Fig 7

6 Conclusions

Although the construction industry is forever searching for new innovations, we cannot disregard traditional skills proven over many years. The evidence of this is clear in the electro chemical repair sector. The inclusion to the industry of computer driven techniques relies on the skill and competence of a sprayed concrete nozzleman to effect a high class repair or overlay. Together, the old and new techniques can take the repair of the construction industry well into the next millennium.

Acknowledgements

Dr Paul Lambert, The Society for Cathodic Protection of Reinforced Concrete
Mr A D Hammond, Electrochemical Techniques for the Repair of Concrete.

6 A DOME FOR YOUR HOME (OR OFFICE) – INNOVATIVE USES OF STEEL FIBRE REINFORCED SHOTCRETE

L.E. Hackman and M.B. Farrell
Ribbon Technology Corp., Ohio, USA

Abstract
In the twenty plus years that steel fibers have been used to reinforce concretes, most of the applications have focused on slab on grade and tunneling applications with shotcrete. Beyond these uses are variations that also call for improved toughness, high flexural or high impact strength. With the premium prices now being paid for wood and steel, concrete structures are increasingly popular. This paper looks at the unique steel fiber reinforced dome headquarters of Ribtec, and how it has held up over the last 14 years.

At the time of Ribtec's initial construction in 1982, steel fiber reinforced shotcrete was used increasingly for slope stabilization, tunneling, and repair of high use areas such as dams. All of these applications have increased significantly in the last 10 plus years, while the more exploratory applications such as dome construction have not taken off. This paper looks at both the successful techniques and the solutions to some of the early problems, which are necessary to make this construction technique more attractive in the years to come.
Keywords: Steel fiber, reinforcement, dome, shotcrete

Introduction

In the early eighties, when looking at expanding its business to a new headquarters location, Ribtec had an opportunity to showcase its steel fiber reinforced technology in its new building construction. The dome construction was chosen for a variety of reasons, not the least of which was the attention that such a unique shape would draw to those looking for more versatile building techniques. Shotcrete was the only option, given the prohibitive cost of forming a dome structure. What made this project so noteworthy at the time was the elimination of structural steel rebar and wire mesh from all but a minimal portion of the construction. The dome would be freestanding and reinforced primarily with steel fiber.

Historically, shotcreted structures are a very small niche even when they are heavily reinforced with rebar. The widespread use of shotcrete in massive placements or thick sections has not yet occurred in many parts of the world, including the United States. This despite the fact that there have been notable successes showing shotcreting as an economical building technique, and not just as a thin section repair technique. The likely situations where these thick shotcrete sections occur are rehabilitation of large structures including seismic retrofit, in construction where framing and placement costs are prohibitive (such as rock formations in zoo habitats), and of course unusual construction shapes such as domes.

Initial Construction

When construction began in March 1982, there was no precedent for a structural dome reinforced with steel fiber that Ribtec could draw from. In order for the local authority to grant a building permit, it was necessary to convince them that such an innovative building practice was structurally sound. This could only happen by allowing us to move away from the specifics of a design specification to a performance specification. This was certainly not easy, but it's likely to be easier today given the increased use of steel fiber reinforced concrete (SFRC). Having secured a permit, the goals were then to demonstrate that it could be done, and at the same time to realize an overall cost savings for Ribtec.

The savings were to come in two areas: the lower labor costs of using steel fiber instead of the conventional mesh or bar, and the reduced volume of material used to achieve the components' structural requirements.

Placed on a 5.2 acre industrial site, the project originally consisted of 22,000 sqft of manufacturing space and 10,000 sqft of office space. Since its completion in December 1982, 10,000 sqft of manufacturing space has been added with another 10,000 sqft of additional manufacturing space expected in the latter part of 1996. The use of SFRC is throughout all phases of construction in the form of precast tilt-up panels, cast-in-place structures, and all floors and pavements. The shotcreted dome still stands as the focal point and the feature that sets the building apart (Figures 1 and 2).

Dome design and construction

The dome shape was achieved by using balloon forms, and applying the shotcrete to the underside of the balloon. After first applying a layer of foam insulation which added rigidity and protection to the balloon, a four inch thickness of shotcrete was layered in. The overall dome design was based on two spherical shells with a base diameter of 100ft.

Figure 1.
Ribtec dome after completion of initial construction in 1982.

Figure 2.
Overall picture of facility including manufacturing area.

One shell was inverted, sitting in a footer 96ft in diameter and extending 12ft above the footer to a 100ft diameter tension ring. The other upper shell extends 20ft above the tension ring as a truncated spherical shell. In order to keep the building light and airy, a 20ft diameter skylight was added to the upper part of the upper shell.

This design flexibility was achieved by using two major compression rings. The footer acts as one compression ring. The other is in the upper shell and surrounds the 20ft diameter skylight. The upper compression ring relieves all the major shell loads from being transferred through the skylights; therefore the sprayed concrete structure is freestanding. The tension ring occurs at the maximum diameter of the dome, approximately 12ft above the footer.

The entire structure is tied to the footer by the use of rebar as seen in Figure 3. The footer was poured 1ft deep, 2ft wide and 4ft below grade to reach soil with sufficient bearing strength to meet the design requirements. This also contained 100lb/cuyd of steel fiber, though the fibers were 2" in length rather than the 1" length fibers used in the shotcrete process.

Large window openings were placed at the main entrance to the building and on either side of the entrance and are shown in Figure 4. These 30ft long openings were spanned by using steel C channels in addition to rebar. In these cases, steel fiber combined well with the rebar to create a fiber concrete composite containing continuous steel.

Figure 3.
Shotcreting base of dome at footer location.

Figure 4.
30ft long window openings on either side of the main entrance.

Shotcrete Mix

The primary drivers for the mix design were pumpability and performance. A wet shotcrete mix was used to accommodate the equipment and the experience that we had to that point. In order to pump a wet mix, we increased our water cement ratio to 0.5 and subsequently ran at a higher slump than we would have had we used the dry shotcrete process. The wet mix shotcreting is easier to handle for a work crew, since dry mix shotcreting requires skillful manipulation by the nozzle man to completely mix the material stream as it exits the nozzle. In the next dome that we built, we used the dry process with improved results and relied more fully on the skill of the nozzle man to place the material. In addition to the improved performance from the lower water cement ratio, we actually had less rebound in the dry mix which is not an expected result when compared to the wet mix. This was perhaps because we had moved well up the experience curve.

As noted in the following mix design, a water reducer was used to give us the desired fluidity and meet our design performance requirement.

	lb/cuyd
Type 1 Cement	846
Fine Aggregate	2440
Water	423
1" Steel Fibers	100
Water Reducer	4oz

Since the floor had already been poured, plastic sheeting was used to cover the floor for protection from the rebound. The amount of rebound is a cost factor both in terms of wasted shotcrete and cleanup. Today's mixes are designed to minimize rebound with additives such as silica fume which tends to make a stickier mix. With SFRC, silica fume additions not exceeding 10% are more common because of pumping considerations. Certainly, we would modify the mix to obtain a balance between reducing rebound while still maintaining pumpability if we were to build a similar dome today.

Barrel-Vault Roof Sections
The roof of the manufacturing area was used to demonstrate the first ever barrel-vault roof design entirely reinforced with steel fiber. Each section measuring 25ft x 100ft was formed in the same manner as the main dome except for a difference in the balloon shape. First a layer of insulating foam to add rigidity was followed by a layering of 2 to 4" of shotcrete. Again the same shotcrete mix was used with 100lb/cuyd of 1" steel fiber. Blockouts were made in the top of the barrel-vault for skylights and exhaust fans.

Some flaring at the bottom of each roof section was designed in to build a strong transition to each adjacent roof section. The buildup of shotcrete at these bases tended to be a minimum of 4" in thickness. Rebar tie-ins sticking up every two feet from cap beams at the top of the building were used to secure the barrel vault roof sections in place. In all, eight sections were used to span the 100ft x 200ft manufacturing area. Steel beams encased in concrete inside the building made up the support structure. The steel beams' design was based on two criteria: 1) that the beams act as a tension element in the overall barrel-vault structure, and 2) that the beams support the concrete before curing.

The internal beam structure was supported by the concrete external structure as well as two internal steel columns. The columns were designed to carry the total roof loads, the weight of the barrel-vaults, and the weight of the support beams. They also were designed for the attachment of a bridge crane structure and to carry roof beams for an additional building on each side of the original building. The addition on the west side of the building was completed in 1992.

Figure 5.
20ft diameter skylight was shotcreted in place by use of blockouts.

Report Card
Structurally, the building has held up very well. The dome has tremendous intrinsic strength due to its shape and composite steel fiber reinforcement. It was probably designed a bit conservatively being the first to rely almost entirely on steel fiber for reinforcement.

In terms of aesthetics, the first reaction we get from many visitors beyond the unusual shape is in regard to the amount of natural light. This is generally not your first thought when considering a domed structure. Credit the versatility of shotcrete and the strength of SFRC for allowing these design considerations to be achieved. As you might expect, there are large open areas and unusually shaped offices with high cavernous ceilings. The center lobby is especially striking, opening into a two story circular atrium complete with a spiral staircase to the second floor. The large skylight at the top of the dome broadcasts natural light throughout the building.

The changes that we would make now that we have lived in it for 14 years relate to further minimizing maintenance, and are a result of not being able to move to the top of the learning curve the first time out.

Figure 6.
Current picture of Ribtec dome, 14 years after initial construction.

The first change would be to get better utilization out of the balloon which was used to shape the dome. These types of balloons are reusable. It's conceivable that the balloon cost could have been spread over three or four other similar buildings to further enhance the economics of this building technique. Since we did not reuse the balloon, we would have gained much better use had we not removed it and left it in place to further protect the dome. A second improvement would be to consider expansion joints in the 100ft long barrel-vault roof sections. The temperature swings in Ohio result in too much movement when you consider how solidly these barrel-vaults were pinned at each end. Both changes would further minimize maintenance.

Conclusion

The importance of innovation and forward thinking engineers is the only way towards continual improvement. There were a number of firsts achieved on this project that are notable, including the free standing steel fiber reinforced dome, the barrel vault roof, steel fiber reinforced tilt-up panels and a 2" thick parking slab for all outgoing freight trucks. This dome construction has now been duplicated in homes and commercial storage structures across the United States. Though the number is not large, the technology advances forward and improvements continue to be made. Structures similar to the barrel-vault roof sections have been used by various State Departments of Transportation to construct free standing culverts, where framing is considered too costly or impractical. As more and more successful projects are reported in conferences such as this, the technology is no longer considered so futuristic.

Today, steel fiber and shotcrete go hand in hand toward versatile construction techniques in tunnels, water channels, slope stabilization and countless repair projects. The more innovative work such as dome construction and other structural applications is pushing the limits of the material forward, and expanding the typical applications to increased acceptance.

References

1. Hackman, L.E. "New facility demonstrates Ohio firm's innovations", *Concrete*, October, 1983, pp 48-52.

2. Warner, James "Understanding Shotcrete - Its Versatility", *Concrete International*, September, 1995, pp 58-64.

3. Warner, James "Understanding Shotcrete - Its Structural Applications", *Concrete International*, October, 1995, pp 55-62.

7 LOW COST LAMINATED SHOTCRETE MARINE STRUCTURES

M.E. Iorns
Consulting Industrial Engineer, West Sacramento, USA

Abstract
Laminating techniques can embed high concentrations of reinforcing in wet-mix shotcrete applied to a floating form made of low-cost oriented strand board (OSB) panels. Common labor assembles the form on the water without the aid of carpenters, sprays the concrete, and embeds preformed reinforcing without using skilled steel workers to tie it in place. Floating formwork eliminates the expense of a casting basin or drydock and towing charges. It also removes all size limits on monolithic construction, so large structures no longer need to be built in segments and incur the complications of later assembly on the water. The system reduces costs to less than half that of concrete cast between double forms in a drydock. Potential applications include floating cities, airports and bridges, immersed vehicular tunnels, breakwaters, bridge piers, and caissons of any diameter and wall thickness. A horizontal floating slipform can extrude monolithic pontoons to any length. A vertical slipform can extrude hollow columns of any diameter and wall thickness down to any depth without the heave and pendular problems that defeated previous attempts to downward slipform offshore.
Keywords: Shotcrete, floating slipforms, offshore structures, laminating, ferrocement.

1 Introduction

A floating concrete structure usually costs more than the same structure ashore even though building on shore requires buying or leasing land and preparing a foundation. This is because a floating structure must be built in an expensive drydock or casting basin that may be located a considerable distance from the site, it may require more reinforcing, and the reinforcing must be protected from corrosion by an extra thick cover of concrete that adds weight and cost.

There is one type of reinforced concrete, however, called ferrocement, that is highly corrosion-resistant and watertight yet may be only 4 mm thick, as in the racing canoes made as class projects by many engineering schools. The ferrocement used for boats, tanks, and as a protective sheath for wood, steel or conventional concrete is usually from 10 to 30 mm thick.

Ferrocement is defined by the American Concrete Institute [1] as " . . . a form of reinforced concrete using closely spaced multiple layers of mesh and/or small diameter rods completely infiltrated with mortar . . . " It was invented in France in 1848 and brought to a high state of development in Italy during and after WW II by P. L. Nervi [2] for boats, ships, pontoons and spans of more than 90 m in the Turin Exhibition Hall and the Rome Sports Palace. Arsham Amerikian, Principal Engineer, US Navy, designed a ferrocement LCT (Landing Craft for Tanks) near the end of WW II that survived a hurricane and was driven at full speed on to hard beaches without damage [3]. All the building methods were so labor-intensive, however, that rising wages in industrial countries relegated ferrocement to buildings like the Sydney Opera House and the Amman Mosque where cost was secondary. In low wage countries, ferrocement is widely used for tanks, housing and boats. China is reported to have over 600 boatyards making thousands of craft annually [4]. A photograph of a 90 m Chinese ferrocement ship designed to carry 500 passengers is shown in [5].

In traditional ferrocement, the mesh is tied to a framework of rods, but experiments at the US Naval Ship Research and Development Center in Maryland [6] showed that rods are very inefficient reinforcing. They are not loaded to take advantage of their strength, their spacing allows regions of unreinforced mortar that add to weight but not to strength, and they act as stress concentrators. A remarkable increase in strength-to-weight ratio was obtained with mesh reinforcing alone. Navy personnel built three experimental high performance 40 knot planing boats in 1970 based on hull specifications used by the three leading commercial ferrocement boat builders at that time. Two used rods and one used an all-mesh laminate. Only the all-mesh hull survived the sea tests. The builders who used rods are no longer in business.

2 Laminated ferrocement (LFC)

Research began in 1963 at the Port of Sacramento in California on ways to reduce labor and eliminate the voids often found in traditional ferrocement. The logical approach was to emulate fiberglass spray production methods in a female mold but substitute steel mesh for glass cloth and cement mortar for polyester resin. Test panels were prepared with multiple layers of various types of mesh and submitted to three different dry-mix (gunite) contractors each of whom claimed his nozzle operator was the best. The panels were cured under water for seven days, cut into strips, and inspected for voids. None was void-free. Even if they had been, the rebound produced in dry-mix shotcrete would have seriously hampered construction in a female mold.

The remedy proved to be to use wet-mix shotcrete, lower the velocity to limit rebound, spray a layer of mortar on the form, and press in strips of mesh with a roller designed for that purpose. More layers of mortar and mesh could be added until the section with its included reinforcing met the design requirements. Welded wire fabric and rebar mat could be placed between any of the mesh layers if additional reinforcing is required.

All the mesh types used by Nervi and others were embedded in 12 mm thick panels and tested for ability to deflect without opening cracks that would leak water, and for ability to absorb impact, a most important requirement for boat hulls. Expanded metal plaster lath was found to be both the cheapest and the best performer. Its strands have a rectangular cross section with a larger specific surface than wire and it is known that the larger the bond area the better concrete can accept strain without cracking.

Expanded metal was investigated earlier at Trinity College, Dublin, Ireland [7], with mixed results, but the Irish lath differed from the heavier laths weighing nearly two kilograms per square meter produced in the UK and the US. Tests on the US lath in California showed that a 12 mm-thick ferrocement section reinforced with four layers of lath could deflect 50 mm in a one-meter span without leaking. Tensile tests revealed, however, that lath was only one-third as strong in the direction of expansion (width) as in the longitudinal direction. This is similar to the strength distribution in plywood veneer and can be compensated for by orienting layers in different directions.

3 Horizontal slipform

Figs. 1 and 2 show construction of a pontoon in a floating slipform at the Port of Sacramento in 1970. The form is simply a box with removable ends made of any water-resistant sheet material. Nine millimeter oriented strand board (OSB) works well and costs less than plywood. OSB panels are bolted together on polyethylene film at the water's edge to form the bottom, sides, and offshore end of the box. As each row of panels is added, the box is pushed or pulled away from the beach. The length of the slipform is determined by the rate of production and the time needed for the concrete to gain enough strength to be left unsupported. Production can start as soon as the first few rows of panels are in place.

OSB costs less than four dollars a square meter in the US. Two layers of polyethylene film, framing, and fasteners add another dollar. The cost of labor to assemble the form on the water varies with wage rates and site conditions, but California experience has

Fig. 1 Floating slipform receiving shotcrete from boom pump.

Fig. 2 Finished pontoon being released from slipform.

shown that labor costs should be no more than the material cost. A slipform for a four lane floating bridge 18 m wide with 5 m high sides would thus cost $280 a meter of length plus the end closures.

4 Vertical slipform.

The advantages of downward slipforming from an offshore floating platform have long been known, but it was thought that high waves would create heave and pendular movement that would preclude conventional slipforming methods. Fig. 3 is a schematic

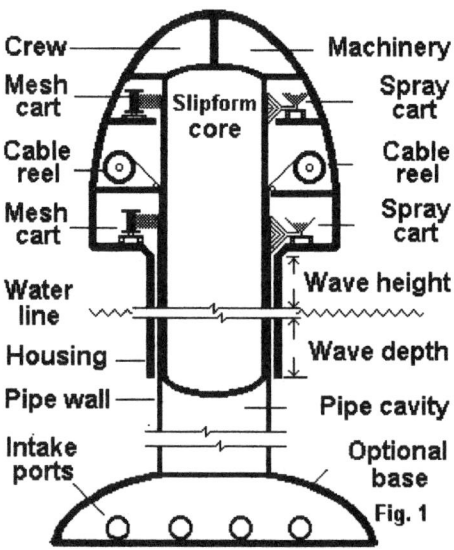

Fig. 3 Floating offshore vertical slipform.

drawing of a patented vertical slipform proposal modeled after the highly stable offshore research vessel "Flip." The same technology used to build LFC water tanks applies multiple layers of mesh and mortar around the slipform core. An enclosure located above the highest waves contains the crew quarters and equipment. Mortar is supplied from a mixer and pump on a service barge alongside. The laminate is protected by the outer housing while it is lowered and gains enough strength for release in calm water below the zone of wave turbulence. Thin-walled pipe would be laminated on the single core shown while thick-walled columns could be made by slipforming inner and outer walls on concentric cores, then filling the space between with a high-capacity concrete pump. Air traps can be built into the walls at appropriate intervals and fed from an air compressor on the surface to regulate buoyancy.

Where high waves are not a problem, downward slipforming can be done on the face of a pontoon built at the site in floating formwork. The production pontoon can be round with a moon pool for producing circular caissons or it can be configured for downward slipforming of bridge piers, ocean mining enclosures, or sheet piling.

5 Applications

Other publications have described LFC's role in building boats [8], wharves [9], breakwaters [10], airports [11], immersed tunnels [12,13], sheathing wood and steel [14], Ocean Thermal Energy Conversion (OTEC)[15], and undersea habitats [16]. This paper will suggest other projects that can be substantially reduced in cost by horizontal and vertical downward slipforming with floating formwork.

Where high waves, strong currents, or navigation requirements preclude a surface bridge, an immersed tunnel with a cross section like that in Fig. 4 with walls one meter thick could be slipformed in floating formwork for about $15,000 a lineal meter and rest on the bottom or float below navigation depth. Henceforth, every proposal to build a high-level bridge should be reexamined to see if an immersed tunnel would cost less.

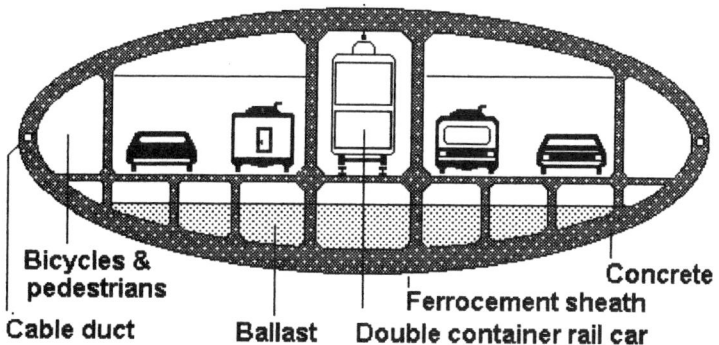

Fig. 4 Cross section of immersed vehicular tunnel.

Floating bridges are a prime application for horizontal slipforming. The floating slipform shown in the photographs (Figs. 1 and 2) could produce a monolithic floating bridge of any length that could be reinforced against lateral current and wind loads. The pontoon shown now supports a floating fueling station in San Francisco Bay, but by adding a little additional deck reinforcement it could become a segment in a floating bridge. It actually cost $43 a square meter of deck area for labor and materials in 1970 and could be built today for less than $100 a square meter. With the additional reinforcing needed to support heavy trucks, a two-lane floating bridge would cost between $1000 and $1500 a meter of length and a four-lane bridge with shoulders and center barrier would cost betwen $3000 and $3500 a meter of length.

Gravity structures of considerable length, width, and depth can be built horizontally in floating formwork, tilted into a nearly vertical position by flooding one end, manuevered into position, and then fully flooded to rest in place on the bottom. Thin pontoons can be used in place of sheet piles to form a seawall and back filled with dredge spoils to create an island or protect a shore line from erosion. Thicker pontoons may be set on edge to form a dike, a levee, or a breakwater with a roadway on top. The seaward face of any floating or fixed structure located where waves are frequent should be configured to convert wave energy into electricity by either the oscillating column or the tapered channel systems that have been successfully demonstrated in Norway.

Large diameter water and sewer pipes are now built in a dewatered trench in short sections that were precast elsewhere. It would be much cheaper and faster to fill the trench with water and make the pipe as a monolith in floating formwork. Outfall sewers could be extruded to any length and depth without the assistance of divers. LFC pipe with a wall reinforced the same as a 20-meter boat hull can be built for less than $50 per meter of diameter per meter of length.

Columns supporting gravity oil drilling platforms and risers for floating platforms, as well as the platforms themselves, could be produced in vertical and horizontal slipforms by a few shotcrete crews at a much lower cost than by present methods. In order to realize these cost savings, owners must specify laminated ferrocement because established contractors are reluctant to adopt a technology that does not utilize their existing plant and equipment.

Offshore moorings can be reduced in cost and their holding power increased by building large plow-shaped concrete anchors in floating formwork near where they are to be placed. They could be made as large as needed for about $100 a tonne. Each anchor would have a hollow buoyancy compartment that would enable it to be towed to its location and flooded to sink it into place. An air hose leading to the surface would permit injection of compressed air to retrieve the anchor for relocation. Orifices could be placed in the anchor to help break suction in soft sediments.

6 Conclusion

Laminating techniques applied to single-surface vertical and horizontal floating slipforms with portable equipment operated by common labor can substantially reduce the cost of marine structures by eliminating skilled carpenters to build double forms, steelworkers to tie reinforcing in place, and the need for a casting basin or shipyard support.

7 References

1. ACI 549.1R-88, "Guide for the Design, Construction, and Repair of Ferrocement," American Concrete Institute, Detroit, 1988, 27 pp.
2. Nervi, P. L., "Structures," W. F. Dodge Co., New York, 1956, 118 pp.
3. MacLeay, F. R., "Thin Wall Concrete Ship Construction," Journal of The American Concrete Institute, Vol. 21, No.3, November 1949, pp. 193-204.
4. Wang Kai-Ming, "The Application of Ferrocement to Boats in China," Proceedings RILEM/ISMES International Symposium on Ferrocement, Bergamo, Italy, July 1981, p.3/47.
5. Iorns, M. E., "A Contractors' Guide to Laminated Concrete," Concrete International, Vol. 11, No. 10, October 1989, pp. 76-80.
6. Disenbacher, A. L,.and Brauer, F. E., "Material Development, Design, Construction, and Evaluation of a Ferrocement Planing Boat, Marine Technology, July 1974, pp. 277-296.
7. Byrne, J .G and Wright, W. "An Investigation of Ferro-cement Using Expanded Metal," Constructional Engineering, December 1961, p. 429.
8. Iorns, M. E., and Watson, L. L., "Ferrocement Boats Reinforced with Expanded Metal," Journal of Ferrocement, Vol. 7, No. 1, July 1977, pp. 9-15.
9. Iorns, M. E.,"Cost Comparisons: Ferrocement and Concrete versus Steel," Concrete International, V. 5, No. 11, Nov. 1983, pp. 45-50.
10. Iorns, M. E., " Breakwaters for Shore Protection, Waste Management, and Wave Energy," Proceedings of Coastal Society 12th Conference, San Antonio, Texas, October 21, 1990, 9pp.
11. Iorns, M. E., "Floating Airport Cost Reduction," Floating Airport San Diego Workshop, University of San Diego, Feb 16, 1990, 11 pp.
12. Iorns, M. E., " Immersed Tunnel Cost Reduction," Proceedings of Third Symposium on Strait Crossings, Alesund, Norway, June 12-15, 1994, 3pp.
13. Iorns, M. E., "New Pontoon Technology Favors All Concrete Construction," Tunnels & Tunneling, London, v. 26, No. 7, July 1994, 3pp.
14. Bowen, G. L., "Sheathing of the Joseph Conrad," Journal of Ferrocement, Vol. No. 6/7, June 1977, pp. 32-34.
15. Iorns, M. E., "OTEC Seawater Pipe Comparisons," International Conference on Ocean Energy Recovery, Honolulu, Nov. 28-30, 1989, 10pp.
16. Karsteter, W. R., et al, "Laminated Concrete for Deep Ocean Construction," Offshore Technology Conference, Houston, May 2-5, 1988, OTC 5634.

8 RECENT DEVELOPMENTS IN MATERIALS AND METHODS FOR SPRAYED CONCRETE LININGS IN SOFT ROCK SHAFTS & TUNNELS

R. Manning
Sprayed Concrete plc, Maidstone, UK

Abstract
In the past 2 years, in the U.K. the NATM method of construction has received considerable attention and criticism. In most cases sprayed concrete is fundamental to the success of the NATM philosophy but sprayed concrete linings stand in their own right. This paper sets out the attributes of using the sprayed concrete process in shafts and tunnel linings, particularly in soft rock. With NATM, because time is of the essence, steel fibres can play a major role in the placement of reinforced sprayed concrete in the primary and secondary linings of tunnels and shafts. To achieve this a structural long life sprayed concrete lining process had to be developed and credibility proved and during the last 2 years a U.K. company Sprayed Concrete Plc has been taking a leading role in a Brite-Euram EC sponsored research and development project. This research is centred on mix designs, admixtures, steel and polypropylene fibres and the evaluation of shotcrete equipment. The research was extensive, considering some 180 mixes. The paper also addresses the importance of proving to designers the credibility of the sprayed concrete process as a long term structural material.
Keywords: Designer confidence, fibres, mix designs, NATM, primary/secondary linings, research.

1 Introduction
The sprayed concrete process has been used for many years for lining hard rock tunnels. Its key attributes are: the time saved by not requiring formwork; high early strength; it forms a freeform tight interface with the substrate. Historically in Continental Europe, hard rock has been "tightened" using concrete thicknesses of between 25mm and 100mm, sometimes incorporating a light mesh which is picked up on rock anchors. The Mid and Southern Europeans have favoured the dry process

and the Scandinavians the wet process, using sodium silicate to stiffen a high slump mix. Where sprayed concrete is used as a primary support to enable a safe working environment, the stop start nature of the tunnelling sequence has attracted the use of the dry process but with careful use of retarders and stiffeners (that need not be chemical), the wet process will in my view lead us into the 21st Century. This trend is now manifest on recent projects in London, but in the majority of underground applications sprayed concrete is only considered as a very useful temporary support system and the process has not achieved recognition as a long term structural tunnel lining.

In soft rock shafts and tunnels an early strength development is essential as the role of the sprayed concrete is to give immediate support. By soft rock, in the United Kingdom, I refer to mudstone, clay, chalk and also in some cases dewatered sand.

With soft rock linings the shell is subjected to immediate load and therefore the concrete material is being stressed and deformed during a very critical period of its life. The most important aspect of any concrete shell is the ability of the concrete to take compressive stress. Reinforcement, encapsulation, day joints and concrete density are also important and must be addressed if longevity is a consideration. From the sprayed concrete practitioner's view point, in situ quality is dependent on mix design, together with the suitability and set up of the key equipment and the expertise in the actual placement. These factors cannot satisfactorily be isolated and must be combined with three fundamental mix design/placement criteria.

1. An intimate bond (contact coat) must be achieved to enable safe build-out on the substrate. If an adequate substrate bond cannot be achieved, then further placements will either result in the collapse of the plastic material or the concrete "hanging" from reinforcement giving rise to poor encapsulation and voiding.
2. The mix must be designed to give an appropriate early strength development, not just to accommodate ground stresses and allow the ground to yield, but to enable the safe progress of the work during the following shift.
3. For permanent support the concrete must reliably achieve the designers long term performance requirement which may be a watertight integrity and also a smooth easily cleaned surface.

2 Convincing the designers

If sprayed concrete is to gain the credibility it deserves and become a major construction material in the 21st Century, it is necessary to examine why designers are wary of sprayed concrete as a long term lining material. If innovation is the state of introducing an invention, to introduce a new idea one needs a platform, such as this Seminar but, most important for our future, we need a willingness by specifiers to listen to and evaluate new ideas.

Like many trade operations, sprayed concrete is often considered to be a specialist activity and in most projects, the main operation is let to a general contractor - in the United Kingdom, the "main contractor" - thus the specialist operates as a sub-contractor. Occasionally the specialist sprayed concrete contractor may receive an enquiry from a consultant at the design stage. But what rarely occurs is a round the table dialogue where the intellectual skills of the designer can integrate with the product knowledge and artisan skills of the specialist contractor to give "buildability". Common sense should dictate that the closer the interface, the better will be the ultimate construction.

The designer usually requires a reinforced concrete section that will reliably take compressive, flexural and tensile forces over a life span dictated by client requirement, that may have to exceed 100 years. The concrete must be strong, dense, encapsulate the reinforcement and be of uniform section. The designer must also have the comfort of being able to use his well tested and tried design formulae.

To some designers it is alarming to be told that formwork is not required, but to tell him he cannot specify reinforcement over 16mm diameter or aggregate over 10mm is often all too much and the sprayed concrete concept is discarded. But it *is* possible to spray material with aggregate sizes larger than 10mm; it *is* possible to achieve a uniform section; it *is* possible to ensure dense homogenous concrete with up to 25mm diameter reinforcement encapsulation; and it *is* possible to achieve a smooth even finish.

The key to the cohesive and plastic properties of the concrete material is the development of the paste, which in the wet process enables transportation of the concrete within the lines without bleeding or logging, and prevents segregation during the spraying process. It is also essential, particularly in tunnelling operations, that the nozzleman works in a dust free environment and can observe and direct the flow of the material around reinforcement. There is a view that, given good air movement and good lighting, the nozzleman will produce good work. My own view is that the mix design, nozzling technique and diligence are far more important, as frequently the mix design can eliminate atmospheric pollution. There is a culture in the UK that limits the quality of the sprayed concrete that probably has roots in mining where sprayed concrete was, and possibly still is, considered an incidental low tech temporary activity carried out by the miners in their spare time or as part of the mineral winning sequence. From a designer's view point it is of course the structural lining that is important, not the mucking out operation which is soon forgotten.

It is imperative that the sprayed concrete specialist has the experience to design and develop a concrete mix that achieves the designer's early strength curve development, while at the same time having the cohesive and plastic properties that enable an appropriate substrate bond to be achieved and remain plastic long enough to be "punched" around say lattice arches and 20mm reinforcing, particularly in areas of laps.

To minimise wastage the material must also absorb the kinetic energy of the next pass as well as remaining cohesive during the projection process, particularly overhead. Where 20mm lattice arches have been used I have seen many cases where the steel has not been fully encapsulated. In my experience this is caused by a lack of plasticity in the mix, over acceleration, atmospheric pollution caused by the spraying process preventing the nozzleman from observing the "cone" of impact and an increasing tendency for nozzlemen to stand 2 metres to 3 metres back, turn up the air and blast the work surface.

3 Primary and secondary lining

The soft rock NATM process relies on the versatility, speed and substrate bond achieved by the sprayed concrete process. Primary and secondary lining describe the shell construction sequence, the primary lining providing initial support. Most designers disregard the primary lining in their calculations and it is generally viewed only as a contractor designed, temporary support system. After convergence monitoring, a second "structural" lining is formed in reinforced concrete usually using static or sliding formwork. Thus the construction effort and expense of perhaps 250mm thickness of reinforced "primary" concrete is ignored.

Some 5 years ago my company developed a structural long life sprayed concrete shaft lining process, constructed top down, using a two pass wet shotcrete process similar to a primary and secondary sequence. To sell the system we had to prove credibility of the total section and to be more economically viable we had to compete on price with precast concrete segments. P.C.C. segments had the advantage that they are cured and Q.A'd before fixing. For the "secondary" lining we decided not to use accelerator and use a basic sprayed concrete mix that could safely achieve 45MPa at 28 days. (Fig. 1) - graph 3. This is a typical structural concrete curve.

Some years later for the Heathrow Baggage Tunnel (*) we offered an accelerated fibre reinforced primary lining to achieve the early strength requirement of a minimum 15MPa in 12 hours. (Fig. 1 - graph 2. The mix safely achieved the requirement with the strength curve levelling out at 12 days. The "primary" lining gives the initial support and the "secondary" lining supplements the first lining as a long term structural entity. When assessing early age strength development it is the shape of the curve that is important. It is essential that when accelerator is used to meet a 12 or 24 hour requirement, the material remains reactive and continues to develop compressive strength. Careful balancing of the reactive ingredients is critical and at Heathrow we were almost caught out by a greater than expected fluctuation in ambient temperatures.

A typical section showing the build sequence of the shaft construction on the Baggage Tunnel is shown in Fig. 2.

4 Advantages

From the designers point of view our sprayed concrete alternative has four advantages:-

1. The concrete is sprayed directly on the earth face achieving a close contact. Therefore the time consuming somewhat hit and miss back grouting process can be omitted.
2. As shaft bottom approaches the tunnel "eyes" can be constructed, which overcomes the hazardous, expensive and time consuming breaking out of segments and in situ concrete making good.
3. The lead-in time is only as long as it takes to design the mix and set up the equipment at the head of the shaft. With P.C.C. segments there is usually a long lead-in period, and extra space required for storing the segments.

* (B.A.A., Bovis, Miller, Beton-und-Monierbau design and construct package).

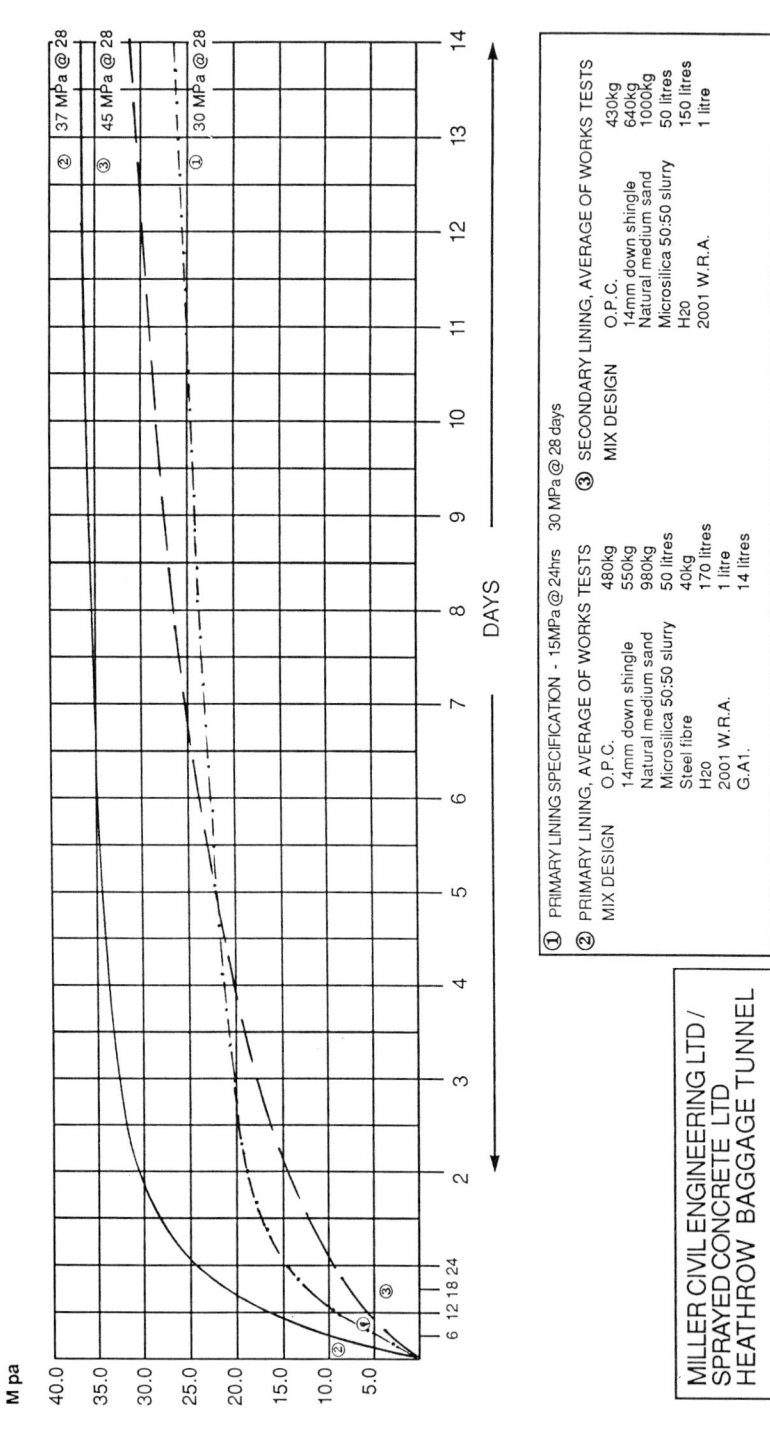

Fig. 1 Compressive Strength Development

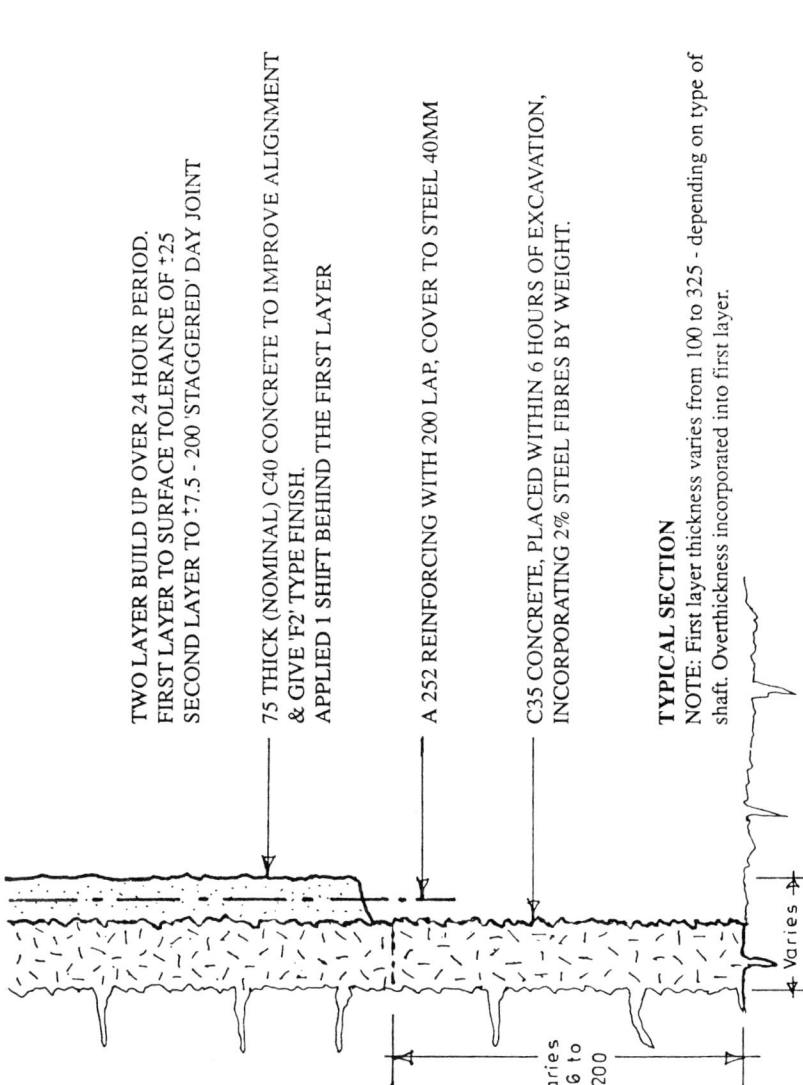

Fig 2 Typical Section: Heathrow Baggage Tunnel - Sprayed Concrete to Shafts

4. Sprayed concrete allows the designer to make daily design changes to incorporate unforeseen ground conditions or changes in section.

The shaft construction follows immediately behind the excavation operation, the first pass using steel fibres. The surface is left as sprayed and during the following shift an accurately aligned steel mesh is installed and the second layer placed. This can either be left as sprayed, screeded to remove any high points, or screeded and rubbed up with a wood float to give an equivalent of the British F2 (formwork) finish. Over the past 2 years a variety of top down temporary and permanent access shafts have been constructed in London clay, up to 65 metres deep using this technique and increasing the section to resist the increasing depth loading. In one project, to minimise outputs, the front alignment was kept vertical and on completion of the sprayed concrete and the construction of the tunnel "eyes", a cosmetic inner lining was slipformed.

5 Research

My company is fortunate to have its own small laboratory where we evaluate concrete mixes and various admixtures, develop spraying nozzles and spraying techniques and also train nozzlemen. We developed our lining system based on **guessing** rather than **assessing** the early age load bearing characteristics of the concrete section. There was fundamental concern as concrete has three states, **fluid, plastic** and **solid** and whereas there is little structural change within the material as it moves from fluid to plastic, when it moves from plastic to solid it goes through its elasto-plastic false set and the complex crystal growth commences. Although it is known that in the early stages the material has a self healing property. Loading too soon may lead to ultimate strength loss and even failure. The problem with soft rock is that the concrete section is subject to immediate loading. It may seem laughable now but in the early stages of development, we considered using a 20mm clayboard void former to take up the possible 48 hour ground movement of the clay to prevent immediate loading and enable the concrete to develop strength. These concerns led us into discussions with Mott MacDonald and Imperial College resulting in a Brite-Euram Research Project. The 4 year Brite-Euram Research Project is part funded with a budget of £1.4m provided by the EC in Brussels to carry out a research aimed at improving sprayed concrete design and practice for underground structures, particularly soft rock and clay with an emphasis on fibres. The Consortium is made up of four British organisations, one Spanish and one Greek.

The research programme has four key objectives;

1. To develop a common understanding and reach an economic balance between the ideal design and what can be reliably achieved by companies skilled in the sprayed concrete process.
2. To evaluate fibres freely available on the European market and their benefits within a tunnelling environment.
3. To evaluate the early age compressive flexural strength development of fibre reinforced sprayed concrete using triaxial testing.

4. To achieve cost control by integrated value engineering and advancing techniques to reduce waste.

This last aim is particularly important when dealing with steel fibres as stainless steel fibres costs per cubic metre can exceed the cost of the base mix. Therefore a wet mix design envelope was presented to the academics who subsequently developed 180 mixes for evaluation. My company brought to the consortium practical expertise and procedures based on materials and climatic experience in Europe and the Middle East. As Sprayed Concrete Plc's core business has been converting designs from formwork and vibration techniques to the sprayed concrete alternative, the company has always had to temper its in-house research and expertise with commercial reality. Thus cost has been an important aspect and has led to mixes based on freely available materials. The same philosophy applies to admixtures. These mixes not only varied the balance of coarse and fine aggregate, but the size of these aggregates and the overall linear gradation. Also considered were the compatibility of cements, admixtures, the plastic properties of the concrete, application characteristics of the concrete, in situ plastic properties of the concrete and physical properties of the concrete and the ultimate material property of the hardened concrete. It is well known that increasing the cement content will compensate less than ideal aggregates that cause difficulty with pumping and segregation but due to concerns over alkalinity associated with high cement mixes, pfa, ggbs and microsilica were also evaluated. It was found that pfa and microsilica slurry would improve workability and the cohesive nature of the mix without significant strength loss. However the argument prevails as to whether pfa and microsilica are pozzolan.

While the practical evaluation was being carried out mathematical models were developed to predict the early age and long term properties of the mix design and these are currently being validated against full scale laboratory tests simulating tunnel linings. Looking at the early age strength we coined a new word - **nascent** - or the at birth strength that will occur within the first 24 hours. By the very nature of the hydration process accelerating the strength development within the first 24 hours will impair later strength development. There has to be trade off, but this is a balancing act that is scientifically and practically achievable.

In the evaluation almost 200m^3 of concrete was sprayed in our laboratory and the 180 mixes were reduced by practical evaluation to 10 mixes that the partners thought represented a major advance and had potential for the European industry at large. Some of the mixes involved 20mm aggregate and 50mm long steel fibres and therefore pumping equipment had to be evaluated and placement nozzles designed to provide the best placement cone and nozzle velocity. A number of surprises arose from the research and the behaviour of the fibres is an example. The steel fibres did not reduce workability as much as expected and surprisingly one 50mm long round fibre caused an increase in slump, probably for two reasons. As this fibre is approximately twice the weight of the other fibres, in a batch of equal weight proportions, there are only 50% of the larger fibre and they are thus more dispersed. Further, the length, round section and cone end of the fibre when acting in the confines of a slump test tend to shear the slumping material. However these fibres did not disperse easily in the mixer and had to be added gradually. If they were added too fast they would "knit" together into a large ball. This problem did not occur with the other types of fibre. Polypropylene fibres did not cause a change in pumpability, probably due to their flexible nature and depending on the make of fibre a reduction in slump can occur. Whereas they can make a mix more cohesive they do not impart any structural benefit. Steel fibres however will make sprayed concrete linings a serious structural option. The appended photographs show how placement techniques and mix designs are fundamental in achieving a dense concrete section.

There is a view that the European sprayed concrete market has been dominated by some chemical manufacturers who have endeavoured to turn sprayed concrete into a chemical cocktail, selling their "wonder" products to the designer/specifiers in the hope of named inclusion in the specification. Many of these companies are connected with equipment manufacturers and thus their grip on the market can be tightened at will. I have long been suspicious of the claims made by manufacturers of sprayed concrete admixtures and we proved that the majority of admixtures that we tested within our programme did not live up to the manufacturers' claims and some were incompatible and inconsistent in their performance from batch to batch. It was soon established that the reactivity of nearly all admixtures, particularly accelerators, were temperature dependent and it would appear that most of the products were developed in a comfortable $20\,^{\circ}C$. My own company works in temperature environments from just above freezing to $28\,^{\circ}C$. Our trials have shown that some accelerators require double dosing below $10\,^{\circ}C$ while others at $28\,^{\circ}C$ cause an immediate set and totally inhibit workability.

Pre-construction trials on site may give some confidence to the specialist contractor and, even the manufacturer, but they do not give long term confidence to the designer. Perhaps this is a fundamental reason why we have problems in achieving credibility for accelerated sprayed concrete as a long term structural material. Furthermore, many mixes are sold dried and bagged as proprietary "high strength", "high bond", "high tech", rather like an in the bag cake mix product where only water and air has to be added. In a few cases these products do have merit but in the main costs to the client are high - and can be 20 times the base material cost. This is more than the sprayed concrete market can stand. If we cannot show a competitive edge, innovation alone will not take sprayed concrete into the 21st Century.

Conclusions

The results of the Brite-Euram research programme will not be published until late 1997. It is already highlighting that with co-operation and understanding between specifier and specialist contractor, and perhaps some value engineering, rebound and wastage can be significantly reduced and lead to substantial cost savings. Overall I believe that the result of the research will lead to a reduction in sprayed concrete costs of 30%, while proving that sprayed concrete is a credible permanent material for tunnel and shaft construction.

To date the research has proved that sprayed concrete **really is** concrete.

Appendix 1

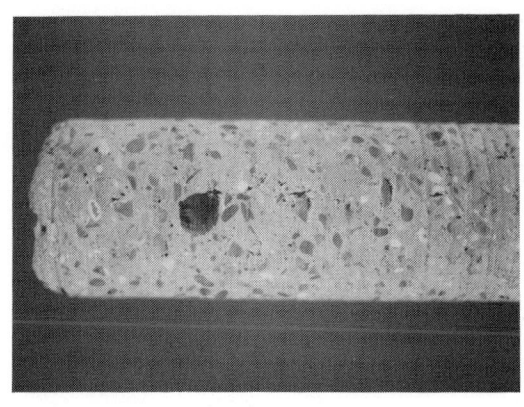

Photograph A - A 100mm core taken from the crown of a NATM tunnel. The 450kg O.P.C. mix incorporated polypropylene fibres at 1kg per m^3. The core was taken in the line of placement and shows poor compaction and fibre "bridging" over the final 75mm. This was caused by a combination of accelerator over dosing and lack of impact velocity - probably in the region of 2 metres per second.

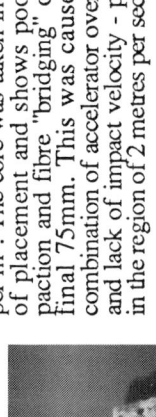

Photograph B - A 100mm core taken from the crown of a 400mm thick rail track arch lining. The mix design: - O.P.C. 400kg, Medium Natural sand 1100kg, 10mm aggregate 650kg, Microsilica 50L, M.B.T. Rheobuild 716 6L, Delvocrete stabiliser 1L SA 145 accelerator 16L.

The core shows a slight shadow behind the front (20mm) reinforcing. Probable cause could be a shortcoming in the nozzle technique or the nozzleman being unable to "shoot" from a wider angle. This core crushed at 54MPa at 28 days.

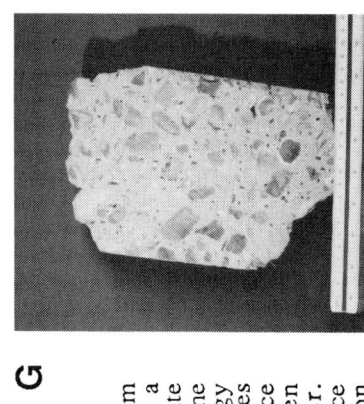

C

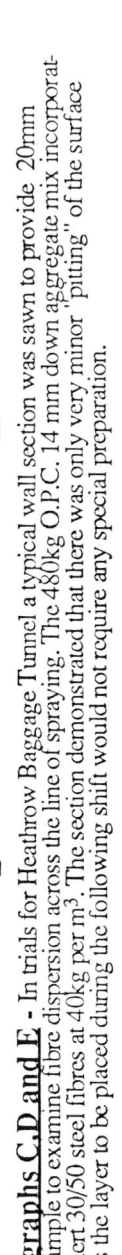

D

E

Photographs C, D and E
In trials for Heathrow Baggage Tunnel a typical wall section was sawn to provide 20mm wafer sample to examine fibre dispersion across the line of spraying. The 480kg O.P.C. 14 mm down aggregate mix incorporated Bekaert 30/50 steel fibres at 40kg per m³. The section demonstrated that there was only very minor "pitting" of the surface and thus the layer to be placed during the following shift would not require any special preparation.

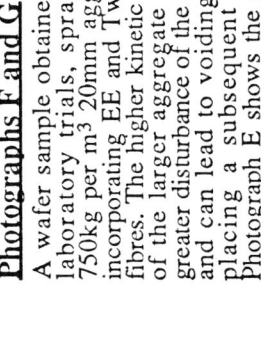

F

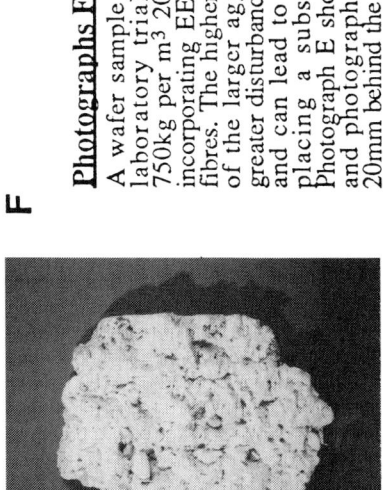

G

Photographs F and G
A wafer sample obtained from laboratory trials, spraying a 750kg per m³ 20mm aggregate incorporating EE and Twincone fibres. The higher kinetic energy of the larger aggregate causes greater disturbance of the surface and can lead to voiding when placing a subsequent layer. Photograph E shows the surface and photograph F the section 20mm behind the surface.

9 SPRAYED CONCRETE FOR THE STRENGTHENING OF MASONRY ARCH BRIDGES

C.H. Peaston, B.S. Choo and N.G. Gong
Dept of Civil Engineering, University of Nottingham, Nottingham, UK

Abstract
An experimental programme was carried out to assess the increase in the ultimate load capacity of circular masonry arches strengthened by a concrete relieving arch. A series of five 2.5 m span brick arches was tested to ultimate load before, and after the application of either a 75 mm or 40 mm layer of un reinforced wet mix sprayed concrete to the arch soffit. The experimental results suggest that the use of the modified MEXE method [1] underestimates the increase in strength. Less conservative guidance is provided by the use of one of a series of finite element programs developed for the analysis of masonry arches [2,3,4]. The finite element technique indicates that the condition of the interface between concrete and masonry is a significant factor affecting performance and leads to upper and lower bound predictive equations corresponding to assumptions of perfect and zero bond. The accuracy of the analytical technique for the specific case of relieving arches increases confidence in the results of a parametric analytical study which examined other arch configurations and repair techniques [5].
Keywords: Bond, finite element analysis, masonry arches, MEXE method, sprayed concrete, strengthening.

1 Introduction

Bridges in the UK are currently being reassessed for heavier vehicles, as defined in the Construction and Use regulations, and are now required to carry 40 tonne vehicles. Local authorities, who are responsible for the reassessment of the bridge stock in response to the increase in vehicle weight, face a choice between strengthening bridges which fail the assessment regulations or putting in place a weight restriction.

Masonry arch bridges, which have proved to be durable and reliable with limited maintenance requirements and are typically over 100 years old, comprise about 40% of the UK bridge stock. The Department of Transport's publications [1] for the assessment of highway bridges and structures indicate that arch barrels may be strengthened by means of a concrete saddle or relieving arch, however there is no guidance for the assessment of the strength of such arch bridges.

This paper describes a series of arch tests conducted to provide experimental data on the load carrying capacity of masonry arches strengthened by a sprayed concrete relieving arch. The work was undertaken as an addition to a continuing programme of masonry arch tests which has the overall aim of identifying the serviceability limit state. After these arches have been tested to ultimate load they are strengthened by the application of a layer of sprayed concrete to the arch soffit and retested.

Comparison of the experimental information was made with two possible means of estimating the increased capacity of the repaired masonry arch. The first of these is to use the modified MEXE method [1] with improved condition and other factors, although this technique cannot account for the composite behaviour of the concrete relieving arch and the brick/masonry arch barrel. The second method involved the use of one of a series of finite element programs which have been developed at the University of Nottingham for the assessment of arch bridge behaviour.

These programs have previously been validated [2,3,4] against large and full scale tests (actual redundant arch bridges) commissioned and/or carried out by the Department of Transport (Transport Research Laboratory) and British Rail Research as well as other small scale model tests. The programs were shown to be capable of accurately predicting the collapse loads of arch bridges subjected to line loads applied across the width of the bridge, provided reasonably accurate geometric information and material properties for the arch barrel and fill are available. In addition, the 2-dimensional finite element program has been shown to provide accurate predictions of crack patterns and deflected profiles of arch bridges with and without ring separation.

The 2-dimensional finite element program has been used in a parametric study of repaired or strengthened bridges [5,6,7] although at the time there was little data with which to provide experimental validation. The test programme described here was carried out in part to provide data for this purpose and therefore to increase confidence in the accuracy of the finite element analysis.

2 Experimental programme

2.1 Test arch details
All the test arches were constructed within the Laboratory of the Department of Civil Engineering at Nottingham University. They were constructed with standard size solid fired clay facing bricks with a compressive strength of about 20 Mpa, and had a span of 2500 mm between concrete abutments which were bolted to the laboratory strong floor. The brick arch barrel was constructed using a timber form and the mortar allowed to cure for 28 days before the form was removed. The arches had a span to rise ratio of 4:1 and a width of 1000 mm. A steel framework supporting plywood panels was then bolted in place to form the spandrel walls which retained a granular fill. There was no connection between the spandrel walls and the arch barrel. The geometry of the test arches is shown in Figure 1.

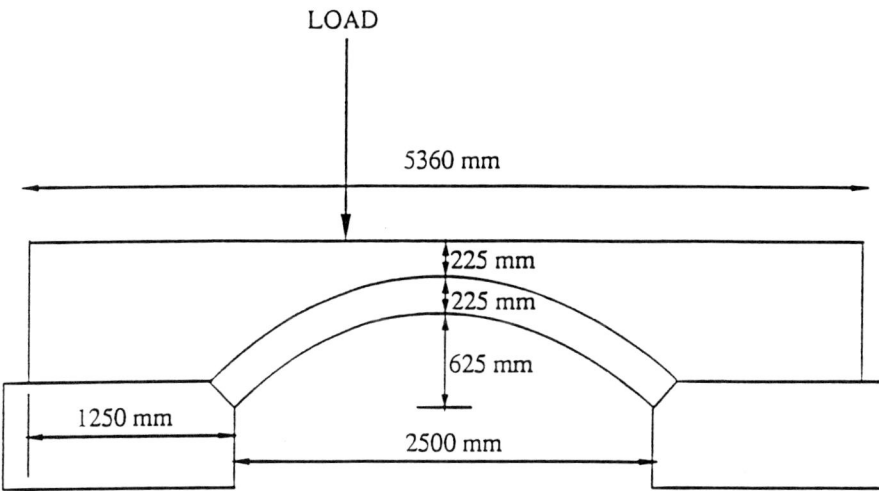

Fig. 1 Test arch - general arrangement

As part of the original experimental programme the first in a series of five test arches was subjected to a monotonic load test to failure (ie to the stage where there was no increase in load carrying capacity). Each of the subsequent arches were first subjected to an elastic load test, which was used in conjunction with load deflection and strain data from the preceeding tests, to predict the ultimate capacity of each arch, and then to varying regimes of cyclic loading. The magnitude of the cyclic load was determined as a percentage of the estimated ultimate capacity with the intention of applying four million load cycles prior to a monotonic test to failure. All loading was applied to the surface of the fill material at the quarter point of the arch span through a single jack located at the centre of a stiff spreader beam of width 500 mm and spanning the full 1000 mm width of the arch.

Details of the estimated capacities and the magnitude of the cyclic loads are given in Table 1. In the case of the fourth arch, failure occurred after a relatively small number of cycles at 80% of the estimated capacity and it was not possible to conduct a test to ultimate load. The fifth arch was accidently overloaded to failure before the cyclic loading regime had commenced and it could not be tested to ultimate load either. A further elastic test lead to a reduction in the estimated ultimate capacity of the damaged arch from 190 kN to 170 kN as indicated in Table 1.

Table 1. Details of arch testing prior to application of sprayed concrete

Arch	Estimated capacity (kN)	Cyclic load (kN)	Conditioning prior to ultimate load test	Ultimate capacity (kN)
1			none	213
2	220	110	4M cycles	237
3	245	135	4M cycles	270
4	200	160	failed at 4,700 cycles	not tested
5	190		accidently overloaded	not tested
	170			

2.2 Sprayed concrete strengthening of test arch

Once testing of the original arches was complete the load was removed and the arch allowed to recover prior to the application of a layer of wet mix sprayed concrete to the arch soffit. At ultimate load the arches were all significantly cracked and some showed signs of ring separation although in all cases they recovered to within a few millimetres of their original profile once the load was removed. Plywood forms were attached to the side of the arch and cut to profile desired of the repair application. The side forms were used as screed rails to check the depth of the application across the width of the arch soffit. The soffit was prepared by needle gunning the brickwork prior to the application of the sprayed concrete to provide as good a bond as possible between the brick arch and the concrete repair. There is evidence that this is not the most effective means of preparing the substrate because microcracking occurs just below the surface [8]. However the use of a potentially more effective method such as grit blasting or water jetting was impracticable within the confines of the laboratory.

The material sprayed was a typical wet mix sprayed mortar based on a mixture of OPC and sand with a smooth particle size distribution varying from a maximum size of 2 mm down to 200 μm. The mix also contained a light weight filler to reduce its density and an acrylic polymer which was used to improve the cohesive properties of the wet mix and improve the bond and durability of the hardened insitu material. The polymer also imparts greater plasticity to the mix. A 3 mm long polypropylene was also used to improve the build characteristics of the wet material as it was being sprayed.

The material was mixed in a pan mixer using the recommended water/powder ratio of 0.12 as a guide although the water content was adjusted by the spraying operatives to obtain a suitable consistency. The wet material was placed in the hopper of an electric Putzknecht S30 worm pump which delivered the material through a 32 mm diameter hose and centre feed nozzle to spray the arch soffit, including the abutments, in a continuous operation. The first two arches were sprayed to a depth of 75 mm and the remaining three were all sprayed with a 40 mm layer. The repaired arches were allowed to cure for a period of six days before conducting a second monotonic load test to failure. The only exception to this was the fifth arch which was tested at 3 days because of a constraint in the project programme.

Samples of the wet material were taken from the pan mixer to make between six and eight 100 mm test cubes and a 450 mm square by 50 mm deep test panel was also sprayed. Four of the cubes were cured in water and four in air under the test arch. They were tested in pairs at the same time as the arch test and at twenty eight days. The test panel was also cured under the arch before being sawn into prisms of nominal 50 mm square cross section for flexural testing at 28 days. Either five or six prisms were tested in four point bending over a 300 mm span in a screw driven test machine at a constant rate of crosshead displacement equal to 0.1 mm/min.

In the case of the first two arches which were sprayed to a thickness of 75 mm the insitu material was sampled after the arch had been tested by taking 63 mm diameter cores. The cores were cured in air and were compression tested in groups of six at 14 and 28 days. A limited number of tensile core pull-off tests were conducted using 50 mm diameter cores on samples of the insitu material from the fifth arch using the 'Limpet' apparatus.

3 Results

3.1 Material properties

Results from the cube, core and prism tests are given in Table 2 which shows that the material properties were fairly consistent from one arch to the next. The exception being the fifth arch in which the material appears to have been stronger than the other arches. The first set of cubes were tested at 3 days and had strengths comparable with the 6 day results from the other arches while the 28 day results exceed those from the other tests. The strength of the air cured cubes was slightly greater than that of their water cured equivalents which is typical of mortar mixes containing a polymer.

The density of the cubes varied between 1890 kg/m^3 and 1985 kg/m^3, the less dense cubes generally having lower strengths, although it is more likely that small variations in the water content explain the variation in the cube strengths. The density of the cores taken from the insitu material varied between 2070 kg/m^3 and 2105 kg/m^3, suggesting that the compactive effort imparted by the spraying process is significant and no doubt contributing to the relatively high 28 day estimated insitu cube strengths when compared with the 28 day cube strengths. The modulus of rupture results are also fairly consistent and are typical of the material.

3.2 Strengthened arch tests

The results of the test on the strengthened arches are given in Table 3. The first two tests give a consistent result of 1.75 for the ratio of the strengthened to the original capacity. The results for the 40 mm application are more variable and depend on what value is taken for the strength of the original arch. The fourth arch failed during cyclic testing at 160 kN which is probably more representative of the strength of the original arch than the estimate of 200 kN. The fifth arch was estimated to have a reduced capacity of 170 kN after it was accidently overloaded. Taking these two values for the respective strengths of the fourth and fifth arches leads to load ratios of 1.48 and 1.38 giving an average load ratio from the three tests with a 40 mm application thickness of 1.40.

The failure modes of the strengthened arches were influenced by the damage sustained during the initial test to ultimate load. The cracks present in the masonry arch opened up again and eventually extended through the concrete layer although none of the arches showed any evidence of debonding at the brick/concrete interface. The pull-off bond tests conducted on the fifth arch all failed at the interface although a thin layer of brick adhered to the pull-off core, suggesting possible microcracking resulting from the method of surface preparation [8]. The average of five pull-off bond tests was 0.76 MPa.

Table 2. Cube, core and prism test results

Arch	6 day cube strength (MPa)		28 day cube strength (MPa)		Estimated insitu cube strength (MPa)		Modulus of rupture (MPa)
	air	water	air	water	14 day	28 day	
1	37.5	36.0	44.5	42.0	51.5	62.5	7.08
2	34.5	32.0	43.0	41.5	47.0	60.5	6.71
3	40.0	37.0	48.0	45.0			6.46
4	41.0	37.5	50.5	47.0			6.63
5	37.5	37.0	53.5	49.5			6.20

Table 3. Results of tests on strengthened arches

Arch	Application thickness (mm)	Original capacity (kN)	Strengthened capacity (kN)	Load ratio
1	75	213	374	1.76
2	75	237	415	1.75
3	40	270	361	1.34
4	40	200 (estimate)	236	1.18
4	40	160	236	1.48
5	40	190 (estimate)	235	1.24
5	40	170 (estimate)	235	1.38

3.3 Comparison with MEXE and finite element methods

When assessing an arch bridge by the modified MEXE method its provisional axle load (PAL) is given by

$$PAL = \frac{740 \, (d + h)^2}{L^{1.3}} \tag{1}$$

where h is the average depth of fill at the crown, d is the arch barrel thickness and L is the span. Using this equation, the relative increase in strength of the repaired arch with respect to the original arch can be shown to be

$$\text{load ratio}_{MEXE} = \frac{PAL_{repaired}}{PAL} = \text{(depth factor)} \times \text{(span factor)} \tag{2}$$

where the depth factor is given by:

$$\text{depth factor} = \left(\frac{kd + h}{d + h}\right)^2 \tag{3}$$

in which k is the ratio of the strengthened arch barrel thickness to the original arch barrel thickness, and the span factor is given by:

$$\text{span factor} = \left(\frac{L}{L_S}\right)^{1.3} \tag{4}$$

in which L_S is the span of the strengthened arch. Within the limits of validity of the modified MEXE method (values of h/d less than unity) the span factor is less significant than the depth factor, although both effects should be accounted for. For the experimental arch geometry these factors lead to the MEXE load ratios given in Table 4. This calculation takes no account of any reduced condition factors that may be appropriate depending on the level of damage sustained during the original ultimate load test.

Details of the finite element programs developed for the analysis of masonry arches have been published elsewhere [2,3,4,6,7]. The 2-dimensional version of the program has been used to conduct a parametric study of the increase in the capacity of arches strengthened either with a concrete saddle or a relieving arch [5,6,7]. Among other parameters the study investigated the effect on circular, parabolic and elliptical arches of variations in the relative thickness of a concrete relieving arch and the bond condition between the concrete and masonry. The variation in the average of the three load ratios obtained for 5, 10 and 15 m span circular arches having respective barrel thicknesses of

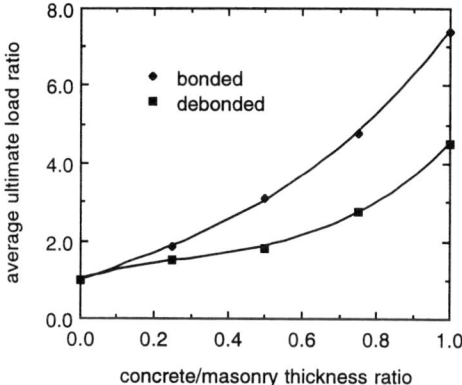

Fig. 2 Average bonded and debonded load ratios

0.33, 0.45 and 0.55 and a span to rise ratio equal to 4, with the ratio of the thickness of an unreinforced relieving arch to the original arch barrel thickness is shown in Figure 2. The analysis was carried out for the two conditions of perfect bond and complete debonding between the concrete repair and the brick arch.

Figure 2 also shows the best fitting polynomials (where LR is the collapse load ratio and t is the ratio of concrete saddle thickness to masonry arch barrel thickness) for the bonded and debonded cases. The equations of these curves are respectively

$$LR = 1.00 + 3.19t + 0.43t^2 + 2.77t^3 \qquad (5)$$

and

$$LR = 1.00 + 2.72t - 4.77t^2 + 5.55t^3 \qquad (6)$$

which represent upper and lower bound load ratios, depending on the condition of the bond between the masonry arch barrel and the concrete. For the experimental arches equations 5 and 6 lead to the upper and lower bound solutions shown in Table 4 allowing comparison with the experimental and MEXE load ratios. The experimentally observed ultimate load ratios both fall within the range suggested by the finite element analysis indicating that equations 5 and 6 provide reliable guidance on the increase in the capacity of masonry arches strengthened with a sprayed concrete relieving arch. Table 4 also provides confirmation that the MEXE method leads to a conservative estimate of the load ratio.

Table 4. Comparison of experimental and analytical ultimate load ratios

Sprayed concrete thickness (mm)	Experimental load ratio	MEXE load ratio	Finite element load ratio	
			Upper bound	Lower bound
75	1.75	1.47	2.20	1.58
40	1.40	1.24	1.75	1.36

4 Conclusions

The load capacity of a masonry arch may be significantly increased through the application of a layer of unreinforced sprayed concrete to the arch soffit that is relatively thin in comparison with the thickness of the arch barrel. The modified MEXE method has been shown to underestimate the improvement in the load carrying capacity. Less conservative guidance, for the specific case of strengthening with an unreinforced sprayed concrete relieving arch, is provided by the application of a 2-dimensional finite element program. The accuracy of the program in this particular application increases confidence in its extension to consider other arch configurations and strengthening methods [5,6,7]. However, in the absence of experimental data on other strengthening techniques caution should be exercised in the extended use of the program.

5 Acknowledgments

The test arches used in this investigation were constructed as part of a programme of research supported by the EPSRC. The materials were supplied by Fosroc International Ltd and the sprayed concrete work was undertaken by Gunform Ltd. The analytical work described in this paper was supported by the Highways Agency (former Department of Transport).

6 References

1. Department of Transport. (1993) *The assessment of highway bridges and structures*, Part 3, BD 21/93; Part 4, BA 16/93.
2. Choo, B.S., Coutie, M.G. and Gong, N.G. (1991) The effects of cracks on the behaviour of masonry arches, in *9th International Brick/Block Masonry Conference*, (ed. DgfM), Vol. 2, Berlin, Germany, pp 948-955.
3. Choo, B.S., Coutie, M.G. and Gong, N.G. (1992) Finite Element Analysis of brick arch bridges with multiple ring separations, in *6th Canadian Masonry Symposium*, Vol. 2, Saskatoon, Canada, pp 789-799.
4. Gong, N.G., Choo, B.S. and Coutie, M.G. (1993) Crack and contact problems in masonry arch bridges, in *5th International Conference on Computing in Civil and Building Engineering*, (ed. L.F. Cohn), Vol. 1, ASCE, California, USA, pp 801-808.
5. Gong, N.G. and Choo, B.S. (1994) *Assessment of masonry arch bridges - strength of arch bridges with concrete saddles*, Department of Civil Engineering, University of Nottingham, Report No SR 94015 to Department of Transport, 32 pp.
6. Choo, B.S., Peaston, C.H. and Gong, N.G. (1995) Repaired Arch Bridges: A Comparison of Analytical and Experimental Data, presented at *Third International Symposium on Computer Methods in Structural Masonry*, Lisbon, Portugal, 10 pp.
7. Choo, B.S., Peaston, C.H. and Gong, N.G. (1995) Relative Strength of Repaired Arch Bridges in *First International Conference on Arch Bridges*, (ed. C. Melbourne), Thomas Telford, Bolton, UK, pp 579-588.
8. Austin, S.A. and Robins, P.J. (eds.) (1995) Sprayed Concrete - Properties, Design and Application, Whittles Publishing, Latheronwheel, UK and McGraw Hill, USA.

10 APPLICATION OF THICK AND HEAVILY REINFORCED SHOTCRETE

J. Warner
Consulting Engineer, California, USA

Abstract
The use of shotcrete for thick and/or heavily reinforced placements has not yet occurred in many parts of the world, and many otherwise well informed shotcrete professionals tend to doubt that such work can be advantageously accomplished. The above not withstanding, thick and heavily reinforced structural shotcrete has been produced for more than fifty years, and procedures are well established to produce quality work in such applications. A mandate to reinforce school and hospital buildings for seismic resistance was made in California, USA following the 1933 Long Beach earthquake. Shotcrete was often used and as practice evolved, ever larger and more heavily reinforced sections were constructed. Placements greater than one meter in thickness and with multiple layers of reinforcing with individual bars as large as 40 mm are now common. Special provisions and techniques required for successful completion of such work are presented. Included is detailed discussion of reinforcing bar layout and jointing, forms and alignment devices that minimize interference with effective nozzling, nozzling technique, provision for effecting proper joints in the work, and quality control.
Keywords: Gunite, reinforced shotcrete, shotcrete, shotcrete application, shotcrete reinforcement, sprayed concrete, structural shotcrete.

1 Introduction

Although shotcrete is most often used in relatively thin and lightly reinforced sections, and some authorities advise that it should not otherwise be used [1], the fact is that a huge amount of both thick and heavily reinforced shotcrete has been produced.
Leadership for such structural use has come from the state of California in the USA as a result of the need for seismic strengthening and repair of earthquake damage. Such application started in response to a mandate to strengthen school and hospital buildings following the 1933 Long Beach earthquake. As the practice evolved, ever larger and more heavily reinforced sections were required and at present it is common

to see the procedure used for placements greater than one meter in thickness, with multiple layers of reinforcing, and individual bars as large as 40 mm [2] [3] [4] [5].

The primary obstacles which must be overcome in heavy structural placements, are obtaining complete filling behind the reinforcement and preventing encasement of rebound, which tends to expand in volume with increasing amounts of reinforcing and growth of section thickness. These obstacles are easily remedied however, through the use of application procedures which are now well established. These include provisions for optimal nozzling through proper layout, installation, and jointing of the reinforcing, avoiding forms or other impediments to nozzling, and provision of joints that resist rebound accumulation. The use of nozzling techniques which minimize rebound accumulation, and appropriate methods for removal of any that does gather as the work proceeds, are also required.

Whereas the procedures to be used are similar for either the dry or wet mix process, use of the wet mixed process which is now highly developed, offers substantial advantages in structural work. Unlike dry mix which requires dispersal of the material as it strikes the receiving surface in order to provide a uniform mixture, the wet mix material is uniformly blended prior to entering the gun or pump. As a result manipulation of the nozzle to complete mixing is not required, and the nozzle can thus be discretely directed through and behind the reinforcing. This results in a dramatic reduction of the amount of rebound, as well as facilitating the filling of isolated areas behind and around the reinforcing.

2 Reinforcing

Congested reinforcing including multiple layers or large bar sizes must be thoughtfully designed, specified, and placed. The particulars will depend upon which shotcrete process is used. Because the requirement for nozzle manipulation to complete mixing in dry mix work precludes discrete nozzle placement, individual bars in multi-layer installations should be offset. Conversely, in use of wet mix which enables discrete nozzle movement, it is usually best to stack the bars of multi-layer installations. This is especially so in the case of bars which are very large, closely spaced, or where many layers exist. Lap splices should be avoided which can be readily accomplished through either welding or use of mechanical splices. Where mechanical splices are used, selection should consider the amount of space they will occupy and only those that are most compact should be used. In extreme cases, the layer of reinforcing nearest the shooting face might be prefabricated but not installed until after the more distant shotcrete has been placed. As with all shotcrete, the reinforcing must be tightly tied and rigidly secured in place.

3 Alignment control and forms

Important fundamental considerations in structural work are to provide the maximum possible latitude in nozzle trajectory in order to facilitate filling around and behind the reinforcing, allow for the unimpeded escape of air and rebound, and minimize the opportunity for rebound to gather. Because of the greater importance of these factors in structural work, special procedures must be used, and in many cases a departure from otherwise common practice will be mandatory.

Although the use forms and/or rigid guides for alignment control is widespread in routine practice, such use in structural applications should be carefully evaluated and generally avoided. This is due to the severe restrictions to nozzling trajectory resulting from their use, and is especially so in situations where supplementary boundary reinforcing exists adjacent to openings or at corners and termini. Additionally, as the depth of the section increases, solid forms greatly inhibit the escape of air and rebound and promote the formation of air voids and entrapment of rebound. For alignment control, ground wires (Fig. 1) which contribute virtually no interference with nozzling, should be used.

Where forms are definitely required such as in construction of a beam with heavy reinforcement on the bottom, sequential erection should be considered. An example would be initial construction of the soffit form only. Following filling of the area under and around the bottom reinforcing which would be facilitated by the ability to shoot from either side as well as from above, a side form could be immediately erected enabling completion of the element.

3.1 Ground wires

Ground wires are usually composed of about 1 mm high tensile wire, and are tightly stretched at corners, offsets, and at intervals of about one meter over planar surfaces. In order to avoid excessive vibration during nozzling and enable the finishers to accurately trim to them, it is crucial that they be accurately positioned and very tightly stretched. Some references call for use of springs or turnbuckles to effect sufficient tension on ground wires [1] [6], however this is not necessary, and skilled ground wire installers routinely perform adequately without such devices.

Where guidance is needed for curved alignments, steel rods about six mm in diameter can be used. These are bent to the correct curvature and securely anchored in place to the proper alignment (Fig. 2).

Fig. 1. Ground wire delineates finish corner but does not restrict nozzle trajectory.

Fig. 2. Small diameter steel rod delineates double curvature contour of bridge pier.

The extensive installation and use of ground wires is fundamental to proper performance, when shotcrete is being used in thick and/or heavily reinforced applications. This requires skilled workers not only for installation of the wires, but working to them as well. Building the shotcrete out to a wire delineating a sharp corner presents a much greater challenge to the nozzleman than shooting to a rigid form or guide. Likewise, trimming back the fresh shotcrete to somewhat flexible wires requires far greater finishing skill than that needed to trim to a rigid guide.

4 Shotcrete application

When encasing large or congested reinforcing, the nozzle must be aimed so as to fill behind the reinforcing bars, and is often positioned at unusual angles to the plane of the work. This of course is contrary to preferred practice of maintaining an angle of impingement of about ninety degrees, but is nonetheless necessary. Also it will usually be held much closer than in traditional work, and will often be held very close to the reinforcing and in some cases will actually penetrate the outer bars (Fig. 3).

The nozzleman must be ever alert to perform in a manner that will minimize the quantity of rebound created, and most importantly, prevent entrapment of that which is. The amount of rebound will vary, but will always grow as the density or congestion of reinforcing increases. In dry mix shotcrete, the rebound quantity can be very high, and within the authors experience has often amounted to more than sixty percent of the material shot on heavily reinforced sections. It is for this reason that most experienced specialists will elect to use the wet mix process in such situations as the quantity of rebound will be significantly lower, usually no more than about ten percent. It should be noted that rebound quantity is greatly influenced by the mix design. Inclusion of silica fume in an amount of eight to ten percent of the mass of the cement will add considerable cohesiveness to the mix and will thus result in a dramatic reduction of rebound.

Fig. 3. The nozzle must be aimed so as to fill behind the reinforcing , will often be held very close to the shooting surface, and sometimes actually penetrate the reinforcing as shown here.

Shooting should start at corners or the ends of sections or other areas that are prone to rebound entrapment and the working face should always slope away therefrom. Thick sections should be built up in layers which slope toward the face at an angle of about 45 degrees, such that any rebound will tend to roll off. In order to minimize rebound entrapment behind horizontal reinforcing, a slope should be maintained in the longitudinal direction as well (Fig. 4). Buildup of the shotcrete in horizontal layers must be avoided, as rebound will tend to become entrapped behind the reinforcing bars as the mass approaches the bottom thereof. The top of the individual placement layers should always terminate between horizontal reinforcing bars in order to provide as clear a path as possible for the rebound to exit.

Regardless of the skill of the nozzleman, as the work progresses rebound will tend to accumulate behind the reinforcing and on the working surface of heavily reinforced or thick sections. It is thus imperative that methods be used which facilitate prompt removal of such deposits, prior to their being encapsulated into the work. Experience has found the best method is to blow any deposits off of the surface with oil free compressed air, immediately as they gather. This is readily accomplished with a blow pipe, which typically consists of standard iron pipe twelve to twenty mm inside diameter, with a valve on one end and the other end flattened to provide a nozzle blast. The "blowman" follows the nozzle movement at all times and promptly blows off any rebound that attempts to gather as illustrated in Fig. 4. Traditionally, the blow man was an apprentice to the nozzleman. With present practice favoring the wet mix process which is extremely physically demanding due to the weight of the solidly filled hose, two qualified nozzleman are often used, alternating between the nozzle and the blowpipe. The importance of using a blowman on applications where rebound is likely to be entrapped, cannot be overemphasized, and in the opinion of the author should be mandatory.

Fig. 4. Thick sections should be built up in layers sloping to the face at an angle of about 45 degrees. The surface should also be sloped longitudinally to resist rebound entrapment.

5 Joints

One of the advantages of shotcrete is its ability to establish monolithic bond between application layers and at construction joints. If such is to occur however, the previously placed joint surface must be properly prepared. All surfaces to which additional shotcrete will be applied must be cleaned of overspray, laitance, or other unsuitable material and roughened. This is easily accomplished by scraping the fresh surface with the sharp edge of a steel trowel, dragging with a bent piece of reinforcing, or brushing with a stiff bristle broom, prior to set. Any formed surfaces or protruding reinforcing steel that has been covered with overspray should also be cleaned prior to the shotcrete hardening.

In order to facilitate the nozzling of subsequent placements and to minimize the risk of rebound gathering as shooting continues, joint locations and configuration should be judiciously selected. As a general rule, joints should be located between reinforcing bars and in areas of minimal reinforcement congestion, to the greatest extent possible. To reduce the risk of entrapping air and rebound in the succeeding work, horizontal joints should be inclined toward the face at an angle of about 45 degrees, (Fig. 5).

A question is sometimes raised as to a possible reduction in load carrying capacity of inclined joints in vertically loaded members and in fact the American Concrete Institute Committee 506, Shotcreting, has recently issued a guide specification which calls for square joints in such members. The fact is however, that literally tens of thousands of such applications have been made over a period of more than fifty years, with no apparent problems. At the same time, a high risk for entrapment of air voids and rebound when square joints are used has been experienced. This risk increases greatly with increasing thickness of the member and/or increasing amounts of reinforcing. The author thus feels strongly that square joints should be prohibited and the long proven practice of inclining all horizontal joints be mandatory in structural applications.

6 Quality control

Because the quality of shotcrete is so dependent upon the skill of the nozzleman and his supporting crew employment of fully qualified and experienced personnel for the work is imperative. Heavily reinforced structural applications require personnel of the highest skill levels and the prequalification of proposed contractors as well as the crew to be used on a project is thus recommended. Experience and competency in more conventional shotcrete application, will not necessarily qualify the workers to adequately perform on heavily reinforced or massive sections. Unless the proposed crew has previously demonstrated satisfactory completion of similar work, full scale test panels or mockups, representative of the most difficult section of the upcoming production work should be required, (Fig. 6). The test panel should be made under job site conditions and utilize the same nozzleman, crew, and equipment that is to be used in the production work.

The test section should be predetermined and detailed in the contract documents. It should be representative of the most difficult conditions which are to be encountered in the actual work, which will usually be that area with the greatest amount of reinforcing steel. Visual inspection for complete filling behind the steel and absence of entrapped voids or rebound can be accomplished by carving into the fresh shotcrete. Following hardening, specimens can be removed by diamond sawing or core drilling for further visual inspection. Should strength be a concern, appropriate laboratory strength tests of the specimens can be made.

Fig. 5. A proper horizontal joint.   Fig. 6. Shooting a full scale mockup.

Where more than one crew or nozzleman is to be used, each should satisfactorily complete test panels. Only those nozzlemen and supporting crews which have been prequalified as a result of satisfactory completion, should be allowed to perform the actual production work. While such qualification tests are generally made on full scale mockups of the pertinent structure, an appropriate section of the actual work can also be used at the beginning of the project.

Small test panels, usually 50 to 100 cm square, which are shot periodically during the production work are often used to evaluate the quality of the in-place shotcrete. Alternatively, "cylinders" made by shooting into wire mesh molds are also sometimes used. It is the considered opinion of the writer that neither of these procedures can verify the existence of quality material in the actual structure. It is much easier for a nozzleman to fill a box with quality shotcrete than to obtain such in the actual work where many variables exist. Similarly, it is difficult to not get a quality specimen when shooting into a mesh basket, as both rebound and excess air are readily expelled. Where it is desired to establish that the mixture is capable of reaching a particular strength, normal cylinder specimens can be made at the mixer in the case of wet mix, and cores can be cut from shot test panels where the dry mix process is used.

Due to the very nature of its composition, shotcrete should not be compared directly to conventional concrete on a micro level. Even very high quality dry mix shotcrete will contain a variation of the water to cement ratio, and all shotcrete will contain some occasional minor imperfections, such as stratification at join lines, small voids or areas of entrapped rebound. Long experience has confirmed that such minor flaws are of little consequence, although the objective should certainly be production of zero defect material.

A qualified inspector should be provided on a full time basis for structural shotcrete work. In addition to constantly observing the nozzling, he should occasionally carve into the fresh shotcrete material visually inspecting areas behind large or congested reinforcing, to confirm complete filling and an absence of rebound or other faulty condition. Obviously, he should also procure required specimens for strength evaluation. In such instances that the strength of the completed work is questioned, core specimens can be procured for testing, as with normal concrete.

Curing

As with any cementitious composition, all shotcrete must be properly cured. Shotcrete almost always has a relatively high cement content and is void of very large aggregate. Even in the wet mix process, the "large" aggregate is generally limited to minus one half inch in size, and the quantity thereof less than thirty percent of the total, resulting in a high shrinkage potential. A minimum of seven days curing is thus mandatory. Continuous wet curing for at least the first three days, longer if practicable, is recommended, followed by application of a liquid membrane curing compound.

Conclusions

Contrary to widely held beliefs, shotcrete can be effectively employed in heavily reinforced and massive placements. Such work has been reliably performed for nearly fifty years. Special techniques are required however, in order to assure complete filling around all reinforcing bars, and prevent entrapment of air and rebound.

Reinforcing layout and splicing must be carefully considered, and performed in a manner which will allow effective nozzling. Forms should not be used around openings or at boundaries, where multiple or large size reinforcing bars exist, or rebound is likely to be entrapped. Alternatively, in order to provide for the greatest possible variation in nozzle trajectory, and allow the easy escape of air and rebound, ground wires should be used for alignment control, and delineation of any offsets or corners. Thick sections should be built up in layers with sloping surfaces in both transverse and longitudinal directions, to encourage the easy escape of rebound. A blowman should be required on all applications, in which entrapment of rebound might occur. The ability of the proposed crew and equipment should be validated through satisfactory completion of full scale test sections under actual job site conditions, prior to the production work, unless their ability has been demonstrated on prior projects. A full time, fully qualified inspector should be provided for all structural shotcrete applications.

6 REFERENCES

1. American Concrete Institute. (1992) *Guide to Shotcrete, ACI 506R-90*. ACI Manual of Concrete Practice, Part 5, American Concrete Institute, Detroit, Michigan.
2. *Chemicals Solidify Beach Sand (1965)* Southwest Builder and Contractor, May 14, 1965.
3. Strand, D. R. (1973) *Earthquake Repairs, Kaiser Hospital, Panorama City, California*, Proceedings, American Society of Civil Engineers, National Structural Engineering Meeting, San Francisco, California, April 9-13.
4. *1931 High School is Retrofit for Seismic Loading* (1979) Engineering News Record, March 1, 1979.
5. Isaak, M. and Zynda, C. (1992) *Innovating with Shotcrete,* Concrete International, Vol. 14, No. 5.
6. Ryan, T.F. (1973) *Gunite - A Handbook for Engineers*, Cement and Concrete Association, London

PART THREE

FIBRES

11 EVOLUTION OF STRENGTH AND TOUGHNESS IN STEEL FIBER REINFORCED SHOTCRETE

A.D. Figueiredo and P.R.L. Helene
Dept of Civil Construction Engineering, Escola Politecnica,
University of Sao Paulo, Brazil

Abstract
This paper presents a brief analysis about the evolution of the strength and toughness of the steel fiber reinforced shotcrete. The strength was evaluated by penetrometers tests along the first ten hours and, after this, it was measured by compressive tests when the stregth was about 8 MPa. The toughness was measured by the ASTM C1018 and JSCE-SF4 methods since the shotcrete presented resistance enough to turn possible to saw the prisms from the test panels. The results shown a strong dependence of the toughness on the matrix characteristics, specially age and fiber content.
Keywords: Age, shotcrete, steel fibers, strength, toughness.

1 Introduction

In Brazil, the tunnels are normally constructed by NATM (**New** Austrian Tunnelling Method) in soft soils, most commonly being clay and sandy clay. This fact provokes a reduction of the stand up time, when compared with that associated with rock underground, and raise up the requirements concerning to the early age strength of the shotcrete lining.

Nowadays, steel fibers have been studied to achieve answers to the designers questions, in the sense of start its applications in Brazil [1]. One of the most frequente question is how the material behave at the early ages? This paper presents a brief analysis about the evolution of the strength and toughness of the steel fiber reinforced shotcrete (SFRS), trying to evaluate how the early age behaviour differs from the 28 days old shotcrete, when it is normally tested.

2 Materials and proportioning

This work presents a simple analysis of the evolution of the strength and toughness of the SFRS. Only one mixture was tested (1:2,82:1,58 - cement:fine agregate:coarse agregate) with the dry mix shotcrete process. The agregate had been proportionated in according to ACI [2] graduation No. 1. The coarse agregate was a 9.5 mm maximum size crushed stone and the fine agregate was an washed river sand. The cement used was a Type III modified with 30% of slag.

A liquid admixture was used to accelerate the shotcrete hardening. The accelerator (potassium aluminate based) content was 2% by weigth of cement and it was added diluted in the water.

The steel fiber used has 25 mm length with hooked ends, and the cross section is rectangular (0.5 mm X 0.4 mm) what provide a aspect ratio equal to 40.5. Two fiber consumption were used: 40 and 60 kilograms per cubic meter of dry-mixture (40 and 60 kg/m^3).

3 Testing procedures

For each fiber content four test panels (50x50x10) cm^3 had been shotcreted in single sequence[3]. Just after the shotcreting a 1 kg sample of the material was colected to determine the final mixture proportion. The real in-situ material proprotion was determinated by washout and sieving method [1][3]. The test panels were involved by a plastic sheet up to 25 hours.

The initial set and compressive strength was estimated up to 10 hours by penetrometers tests. Two types of penetrometer were used: the Meynadier Needle and the Constant Energy Penetrometer [1][4].

When the shotcrete was 5 hours old a test panel was removed to saw the prismatic specimens (10x10x40) cm^3. These specimens were submitted to the ASTM C1018 "Standard Test Method for Flexural Toughness and First Crack Strength of Fiber-Reinforced Concrete (Using Beam with Third Point Loading)" [5] and the JSCE-SF4 Japanese Method [6] when they were 6 hours old and more. The deflection was measured by a "yoke" system [7] and the load-deflection curves were obtained by a X-Y plotter. The edges of the tested specimens were subjected to compression tests in sequence. Both tests were made in parallel direction of spraying.

4 Results

The actual mixture proportion in the test panels were 1:3,37:0,39 (cement:total agregates:water) when the fiber consumption were 40 kg/m^3 and 1:3.53:0.46 when the fiber consumption were 60 kg/m^3.

The compressive strength results evaluated by the penetrometers [1][4] and obtained by the direct compressive tests up to 10 hours are shown in Fig. 1. The average results and their standard deviations are presented in Table 1. with the modulus of rupture. In the Tabel 2. the results concerning to the toughness indices (ASTM C1018) and toughness factors (JSCE-SF4) are presented.

The compressive strength results, obtained by the penetrometer tests and by direct compressive tests, were very close. This fact corroborates the penetrometers application because the non-destructive tests are more suitable for large scale and on-site aplications. Moreover, this kind of tests permits an evaluation of the shotcrete homogenity during the shotcreting at the tunnel front, where the risks of accidents are higher.

The correlation relationship between the average compressive strength, determined in prism (Table 1.) and the age of the SFRS was studied by the least-squares method. The mathematical expression used to try to fit the experimental points was:

$$fc = \frac{A}{\frac{t^{-\frac{1}{2}}}{B}}$$

Where:
fc = compressive strength (MPa)
A and B = constants
t = time (days)

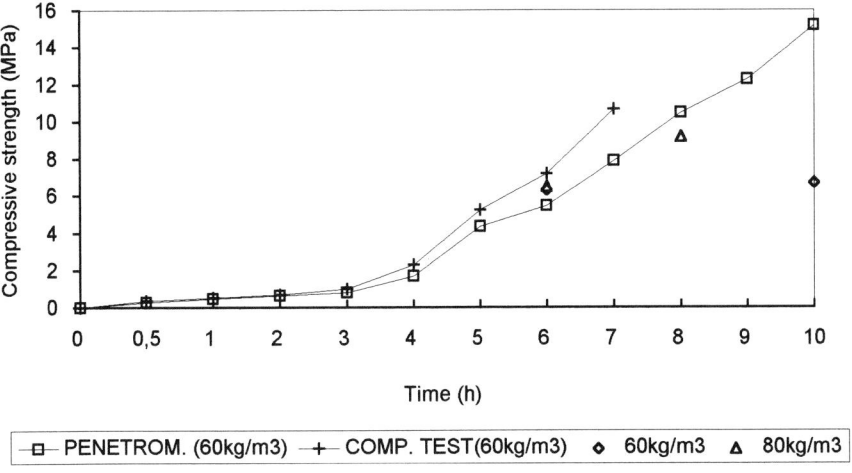

Fig. 1. Compressive strength of the SFRS up to 10 hours.

The expression obtained to the fiber consumption of 40 kg/m³ was:

$$fc = \frac{41.8}{\frac{t^{-\frac{1}{2}}}{2.69}}$$

And the fiber content of 60 kg/m³ of dry mixture shotcrete was:

$$fc = \frac{44.2}{\frac{t^{-\frac{1}{2}}}{2.45}}$$

Table 1. Modulus of rupture and compressive strength obtained by prismatic specimens.

Age (days)	Fiber consumption			
	40kg/m^3		60kg/m^3	
	Compressive strength (MPa)	Modulus of rupture (MPa)	Compressive strength (MPa)	Modulus of rupture (MPa)
0.25	6.3±1.1	1.14±0.03	6.5±0.5	2.05±0.04
0.33			9.2±0.4	2.08±0.34
0.42	6.7±0.3	1.72±0.12		
1	21.2±2.9	4.71±0.22	21.5±1.5	5.07±1.11
4	26.3±4.3	5.72±0.56		
5			28.9±1.8	5.79±0.35
7	28.3±4.6	5.61±0.67		
28	31.2±3.0	6.15±0.87	35.0±1.2	8.91±0.77

Table 2. Toughness indices (ASTM C1018) and toughness factors (JSCE-SF4).

Fiber consumption (kg/m^3)	Age (days)	ASTM C1018 Indices			JSCE-SF4 Factor (MPa)
		I_5	I_{10}	I_{30}	
40	0.25	3.0±0.3	5.2±0.1	13.5±1.2	0.6±0.1
	0.42	2.7±0.1	4.3±0.5	10.2±2.9	0.6±0.1
	1	2.6±0.4	4.4±0.8	11.5±2.1	2.0±0.3
	4	2.3±0.1	3.9±0.1	8.2±0.2	1.2±0.2
	7	2.4±0.2	3.9±0.2	9.7±0.4	1.8±0.2
	28	1.9±0.1	2.9±0.4	5.4±0.1	0.9±0.1
60	0.25	2.9±0.1	4.4±0.1	9.7±0.5	0.6±0.1
	0.33	2.8±0.1	4.7±0.3	11.5±1.2	0.8±0.1
	1	3.6±0.2	6.1±0.5	14.3±1.5	2.1±0.4
	5	3.2±0.5	5.5±0.9	13.3±2.9	2.6±0.4
	28	2.7±0.2	4.8±0.3	11.5±1.0	3.6±0.3

Both mathematical expressions obtained leads to a good fit ($r^2 = 0.92$ for 40 kg/m^3 and $r^2 = 0.96$ for 60 kg/m^3). The little differences between the expressions may be caused by the fact that the two test panels series are made in two different days. Little differences in the equipment adjust, like compressive air pressure, and in the environment conditions may had caused changes in the shotcreting.

The same approach was used on the modulus of rupture analysis. The expressions obtained by the least-squares method were:

$$MOR = \frac{8.5}{t^{-\frac{1}{2}} \cdot 2.5} \quad (r^2 = 0.86) \text{ for 40 kg/m}^3 \text{ and}$$

$$MOR = \frac{9.6}{t^{-\frac{1}{2}} \cdot 2.2} \quad (r^2 = 0.86) \text{ for 60 kg/m}^3$$

Where:
MOR = modulus of rupture (MPa)
t = time (days).

The modulus of rupture results were very similar to the compressive strength. Even the differences may be credited to the fact of the different days of test panels shotcreting, because both fiber contents are under the critic volume of fiber (1%) and they do not influence the tensile strength of the matrix [8].

The toughness indices (ASTM C1018) obtained for a fiber consumption of 40 kg/m^3 shown a constant diminution on the indices values. On the other hand, the consumption of 60 kg/m^3 shown a different behaviour: the toughness indices increased up to one day and decreased after this age. Similar behaviour was presented by the toughness factor (JSCE-SF4) for the fiber consumption of 40 kg/m^3. And, in a different way, the toughness factor increased as the SFRS became older for the fiber consumption of 60 kg/m^3.

The toughness indices obtained for the lowest consumption may be explained by the fact that the "residual strength" provided by the bridge function of the fiber in the post crack deflection increase at a low rate when compared to the first crack strength (modulus of rupture). This is shown by the variation of the toughness factors (JSCE-SF4) compared to the ASTM indices. The resilient energy increases more than the tenacity. So, as the toughness indices are obtained by dividing the area up to a specific deflection by the area up to the first crack, the division result decrease necessarily. In other words, the shotcrete with a low fiber content presents a behaviour most similar to the perfect elastic-plastic material at the low ages or strength.

One of the explainings about this behaviour is the possibility of the tension transfer from the matrix to the fiber may provoke a more intense sliding effect between matrix and fiber, whose intensity is bigger when the fiber content is low. When a higher consumption of fiber was used a better performance was obtained at the post crack deflection. By this fact, a large area under the load-deflection curve was obtained. This fact provoked a constant increase on the toughness factor and postpone the toughness indices reduction, what only happened after one day.

5 Final comments

This work is not conclusive and further studies where necessary, using other types and consumptions of fibers. However the results shown that the penetrometers are very suitable to control the compressive strength at the first hours. The evolution of the compressive strength and modulus of rupture was quite similar for SFRS and plain concrete. On the other hand, the control of SFRS toughness at the age of 28 days is not enough to explain the behaviour at other ages, specially the earliest ones when, in some cases like tunnelling in soft soils, are much more important.

6 References

1. Figueiredo, A.D., Helene, P.R.L. and Agopyan, V. (1995) *Fiber Reinforced Shotcrete for Tunnelling for NTM Brazilian Conditions.* Fiber Reinforced Concrete -

Modern Developments. Ed. N. Banthia and S. Mindess. The University of British Columbia, Vancouver, Canada. 10pp.
2. ACI Committee 506 (1990) *Guide to Shotcrete*. American Concrete Institute (ACI 506R), Detroit, USA, 41pp.
3. Armelin, H.S. (1992) *Steel Fibre Reinforced Dry-Mix Shotcrete - A Study of Fibre Orientation and its Effects on Mechanical Properties*. In: RILEM 4th Symposium on Fibre Reinforced Cement and Concrete. Sheffield, England. 14pp.
4. Prudêncio Jr. L.R. (1991) *Strength Evaluation for Early Age Shotcrete*. In: ACI International Conference onEvaluation and Rehabilitation of Concrete Structures and Innovations in Design. Hong Kong, China. 15pp.
5. ASTM C1018 (1989), *Standard Test Method for Flexural Toughness and First-Crack Strength of Fiber Reinforced Concrete (Using Beam With Third-Point Loading)*, 1991 Book of ASRM Standards, Part 04.02, ASTM, Philadelphia, pp507-513.
6. JSCE-SF4 (1984) *Method of Test for Flexural Strength and Flexural Toughness of Steel Fiber Reinforced Concrete*, Concrete Library of Japan Society of Civil Engineers, 3, pp58-61.
7. Banthia, N. and Trottier, J-F. (1995) *Concrete Reinforced with Deformed Steel Fibers. Part II: Toughness Characterization*. ACI Materials Journal, V.92, No. 2, March-April. Detroit, USA. pp.146-154.
8. Shah, S.P. (1991) *Do Fibers Improve the Tensile Strength of Concrete?* In: First Canadian University-Industry Workshop on Fibre Reinforced Concrete. Proceedings PP10-30. Quebec, Canada.

12 FLEXURAL STRENGTH MODELLING OF STEEL FIBRE REINFORCED SPRAYED CONCRETE

P.J. Robins, S.A. Austin and P.A. Jones
Dept of Civil and Building Engineering, Loughborough
University of Technology, Loughborough, UK

Abstract
The paper outlines an analytical model for predicting the flexural response of steel fibre reinforced sprayed concrete (SFRC). This model is based on a stress-block diagram approach to represent the stress distribution across a propagating crack under increasing flexural load. An experimental testing programme is being undertaken with a wet process steel fibre reinforced concrete mix, including single fibre pull-out tests, flexural tests, compression tests and strain analysis tests, in order to develop relationships which can then be used in the model to predict the stress-block diagram for any given beam deflection. The single fibre pull-out test is described in which the fibre is embedded and then pulled from an uncracked matrix. The test allows a wide range of fibre orientations and embedded lengths to be investigated. The implications of the results are discussed, both for the model and more general for the testing of SFRC and its design.
Keywords: flexural strength, modelling, pull-out tests, sprayed concrete, steel fibres, wet process.

1 Introduction

The inclusion of short, randomly distributed steel fibres into a sprayed concrete matrix has long been recognised as an effective means of improving the mechanical strength characteristics of this material. The resulting composite has greatly enhanced post-crack toughness, fracture and impact resistance. The use of fibres in sprayed concrete applications is particularly attractive in concrete repair and rock support work, where conventional means of reinforcement, such as wire mesh, can be difficult and expensive to install and may often not be as effective.

Dry process sprayed concrete has until now been the dominating method for this type of work in the UK. However, in the future sprayed fibre reinforced concrete (SFRC) using the wet process will become more and more popular as this method offers significant advantages over the dry process. These advantages include: a better working environment, higher and more consistent quality, a controlled and known water dosage, and much less wastage (i.e. rebound).

With recent, and continued, improvements in materials and spraying technology SFRC has the potential to become a more prominent construction method in the future. However its continued development has been hindered by a general lack of confidence in its use as a permanent structural material, which has resulted from the following: an incomplete understanding of the fibre/matrix reinforcing mechanisms and failure process; and a lack of appropriate 'user friendly' design models and procedures, using conventional (and widely accepted) material parameters.

In an attempt to address these points, an analytical model is presented to predict the flexural response of SFRC. An on-going experimental investigation is currently being undertaken to establish appropriate data for use in the model. The investigation uses a typical wet process steel fibre reinforced sprayed concrete mix, and tests cast (as opposed sprayed) specimens so that the material parameters under investigation can be better controlled.

In this paper, the basic concepts of the proposed model are presented together with details of the single fibre pull-out tests. Details and results of the other tests undertaken as part of the whole investigation will be presented in subsequent publications.

2 The proposed model for predicting flexural response

2.1 Some previous models
Previous attempts at modelling the flexural strength of fibre reinforced concrete have usually either adopted an over-simplified approach [1], or have incorporated complex fracture mechanics theory [2,3]. Furthermore most models have usually only been applicable to smooth fibres. However, they all recognise that the key parameter in a fibre/matrix model is the relationship between the post-crack stress response and the corresponding displacement, which is primarily a function of the pull-out behaviour of the fibres.

2.2 Basic concepts of the proposed model
An idealised representation of a cracked SFRC beam under flexural loading is illustrated in Figure 1. The stresses (and resultant forces) over the beam depth can be represented by three zones: compression; uncracked tension; and cracked tension. The cracked tension zone can be further represented by three sub-zones [2]:
- an aggregate bridging zone (resulting from microcracking which then initiates fibre/matrix debonding);
- a fibre bridging zone (fibres fully debonded and partially pulled out from matrix);
- a traction free zone (fibres fully debonded and completely pulled out from matrix)

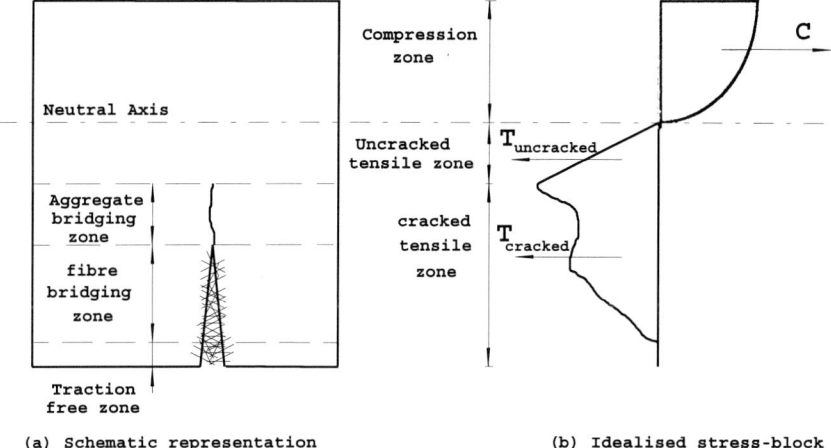

Fig.1, Schematic representation of a cracked SFRC beam under flexural loading

It is clear that the stress-block diagram constantly changes as the crack propagates through the depth of the beam, and that the flexural capacity of the beam at any given crack length is related to: the compressive stress distribution, the elastic tensile stress distribution, the fibre pull-out stress distribution, the position of the neutral axis, and the crack-width. The fibre pull-out stress distribution is the principal parameter which dictates the post-crack response of the beam. If the shape and magnitude of the stress-block diagram can be ascertained then it can be used to calculate the flexural capacity of the beam in a similar way that stress-blocks are used in reinforced concrete design.

2.3 Experimental investigation to obtain data for the proposed model
The experimental investigation primarily attempts to model the flexural response of a typical wet process sprayed concrete mix, reinforced with 30mm long/0.5mm diameter deformed-end steel fibres in concentrations of 0, 0.5, 1.0, 1.5 and 2.0% by volume. This particular fibre type was adopted in the investigation as being typical of the fibres used in a large proportion of wet process SFRC applications. Four tests are being used to obtain the necessary data for the proposed model: single fibre pull-out tests, flexural and strain analysis tests on notched and un-notched beams, and compression tests. This paper discusses the pull-out test, which is being used to obtain the fibre pull-out stress/crack-width distribution for inclusion in the model.

3 Single fibre pull-out test program

3.1 Background
Earlier research has attempted to relate the behaviour of fibre reinforced composites to single fibre pull-out tests [3,5]. However, to the authors' knowledge, all the single fibre pull-out tests reported to date have involved fibre pull-out across an artificial

crack. Consequently, no account is taken of the pre-crack fibre/matrix interface stress distribution or the change in this distribution which occurs at the instant the matrix cracks. In an attempt to overcome this shortcoming, a single fibre pull-out test has been developed which casts the fibre in an uncracked matrix and therefore provides the complete load/pull-out response.

3.2 Materials and mix design

The concrete mix used in the pull-out tests, was chosen as being typical of current wet process SFRC mixes, following a review of published mix designs and the results of mixing, pumping and spraying trials in the laboratory. The mix composition is detailed in Table 1.

50mm long/0.5mm diameter deformed end steel fibres were used in the pull-out tests. The length of theses fibres allows the fibre embedded length to be varied so as to model that expected across a flexural crack (i.e. 0-15mm for 30mm long/0.5mm diameter fibres), but still allow a much greater fibre anchorage in the other half of the specimen so that the fibre only pulls out from one side.

3.3 Specimen preparation and testing procedure

The configuration of a typical test specimen is shown in Figure 2.

Encasement of the fibre within the uncracked matrix was achieved by casting a brass insert, with a 10mm hole located at its centre, at the mid-length of each specimen. This hole served two purposes: to provide continuity of the specimen across the insert, and to form a reduced section at which the specimen would crack during the test. The size of the hole was determined from a relationship proposed by Krenchel [4] to predict the average fibre spacing in a random 3-D fibre composite. To prevent a bond forming between the insert and the specimen, a thin layer of PTFE spray was applied to one side of the insert.

Specimens were cast in two stages within a purpose made mould. Extreme care was exercised to minimise mix variability by adhering strictly to procedures developed for the batching, mixing, placing and curing of the specimens. Firstly, the half of the specimen containing the length of fibre under investigation was cast, using the typical wet-process mix quantities of Table 1. This was left to cure for 24 hours, and the second half was then cast using a rapid setting high strength mortar to securely anchor the fibre in that half of the specimen. To ensure continuity of the specimen across the insert, the specimens were joined by a construction joint located 10mm to one side of the insert (see Figure 2). Once hardened the completed specimens were carefully demoulded and placed in a curing tank until the time of testing. Specimens were tested at an age of 28 days.

Table 1. Wet process sprayed fibre reinforced concrete mix composition (plain matrix)

Cement (OPC)	Aggregate 0-5mm River Gravel	Water	Silica Fume Slurry	Super-plasticiser	Slump	28 day Compressive strength	Hardened density
(kg/m3)	(kg/m3)	(kg/m3)	(kg/m3)	(kg/m3)	(mm)	(MPa)	(kg/m3)
490	1510	245	50	8.1	125 +/-25	65 +/-5	2230 +/-20

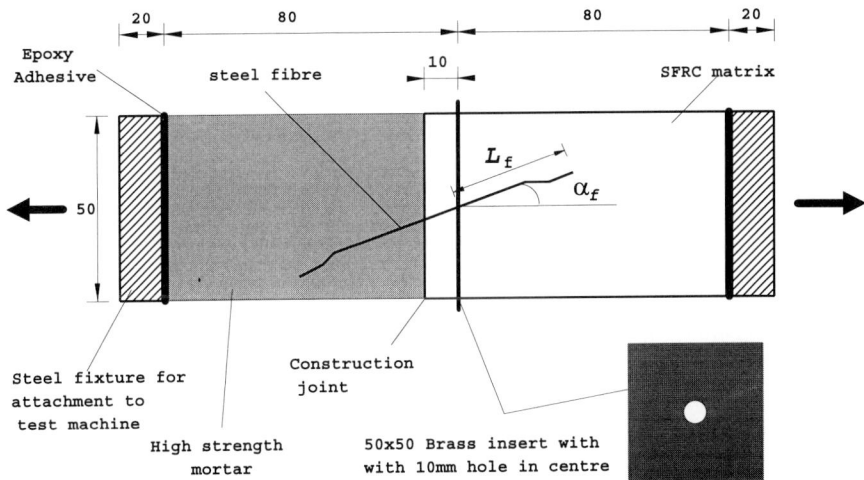

Fig. 2, Configuration of a typical pull-out specimen (all dimensions in mm)

A specially designed block was used to position the fibre accurately at the centre of the specimen during the first stage of casting, and to obtain the correct fibre orientation and embedded length. Fibres were orientated with respect to the direction of loading at 0, 10, 20, 30, 40, 50 and 60 degrees (angle α_f in Figure 2), and at embedded lengths of 5, 10, 15 and 20mm (length L_f in Figure 2). Four identical specimens were cast for each combination of embedded length and orientation.

Pull-out tests were performed with an Instron 6025 screw driven testing machine using a 10kN load cell. Specimen deformation, and crack widths, were measured over a gauge length of 85mm using four LVDT's mounted on a purpose made aluminium frame. One LVDT was located on each side of the specimen to enable an average deformation reading. Load and deformations were digitally recorded using a data acquisition system operating at a frequency of 0.2Hz.

Load was applied at a cross-head travel rate, as follows:

- 0.028mm/min over the elastic region up to the matrix fracture point;
- 0.082mm/min over the initial pull-out region up to a 1.2mm crack width; and
- 1.0mm/min between 1.2mm crack width and complete fibre pull-out.

Pull-out rates up to a 1.2mm crack-width were calculated to be equivalent of the strains developed in the extreme fibre of a 500x100x75mm beam, when tested over a span of 450mm, in third-point loading, with a mid-span deflection of 0.1mm/min (based on EFNARC [6]). 1.2mm was observed as being typical of the crack developed at the extreme tensile face of the beam at the I30 toughness index deflection limit. Beyond this crack-width, the response of the pull-out specimens was considered to be of minor importance to the model, and therefore a nominal pull-out rate was adopted.

4 Results and discussion

Results from pull-out tests using combinations of fibre embedded lengths of 5mm and 10mm, and orientations of 0-50 degrees are presented in Figures 3, 4 and 5.

A typical pull-out load/crack-width curve is shown in Figure 3. The shape of the curve can be characterised by four principal points (labelled A to D on Figure 3.): matrix fracture point (A); initial post-fracture fibre load point (B); post-fracture peak-load point (C); and complete fibre pull-out point (D). It can be seen that immediately after the matrix fracture point, the load does not drop to zero (an assumption usually made in pull-out tests performed across an artificially cracked matrix) but reduces initially to a fibre 'pre-stress', which is equivalent to the stress in the fibre at the instant the matrix fractures. It is the occurrence of this stress which indicates that this pull-out test can account for the pre-fracture fibre/matrix stress distribution, and hence be more realistic of the fibre/matrix response in flexure.

The total post-fracture toughness of individual fibres in relation to their complete pull-out crack-width are shown in Figure 4. Post-fracture toughness was obtained by calculating the area under each pull-out load/crack-width curve between points B and D (as indicated in Figure 3). In general Figure 4 shows, as expected, that the toughness increases as the total pull-out length increases but that the relationship is non-linear. It appears that the end deformation, typically located at a distance of between 2.5-4.0mm from the end of the fibre significantly increases post-fracture toughness. However, fibre orientation has significantly less influence on total post-fracture toughness (when measured to complete fibre pull-out).

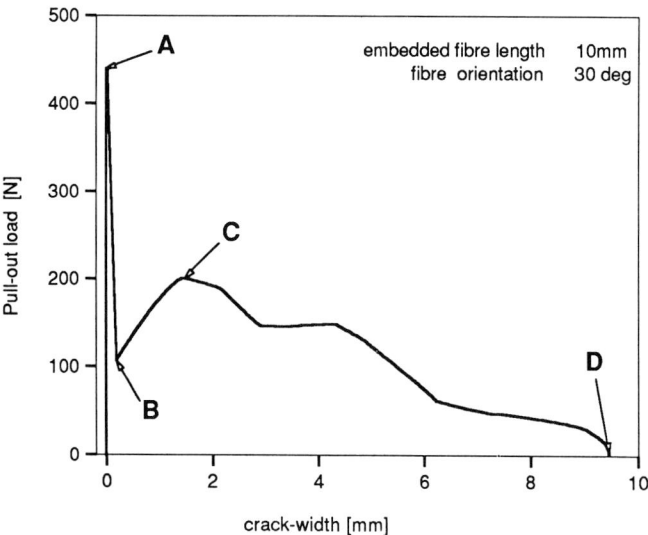

Fig.3, Typical pull-out load v's crack-width response

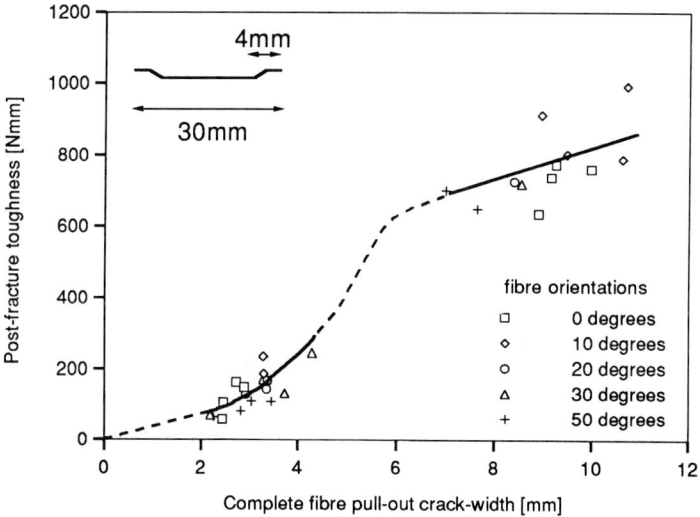

Fig.4, Total post-fracture toughness of individual fibres v's. complete fibre pull-out

The variability in the total pull-out lengths for the 5mm and 10mm fibres shown in Figure 4, 2.4-4.3mm and 7.0-10.7mm respectively, is caused by the position and profile of the matrix fracture interface relative to the embedded fibre end. Although all the specimens fractured at the reduced section, the exact position and profile of the fracture was rarely coincident with the face of the brass insert (the point from which embedded lengths were measured during specimen casting).

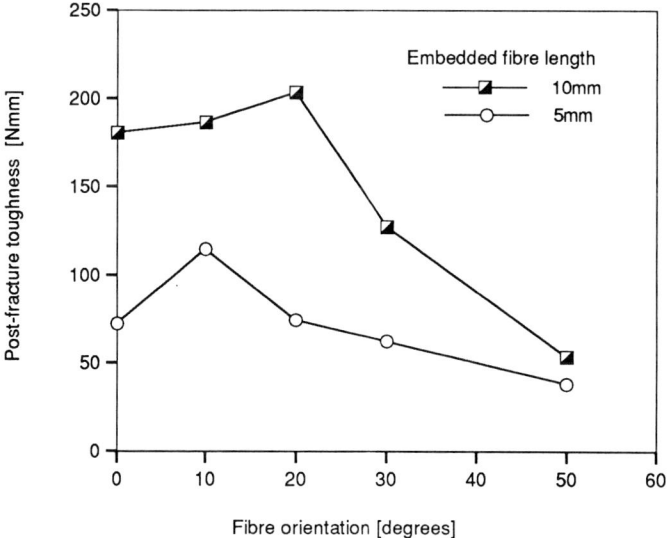

Fig.5, Average post-fracture toughness up to crack-width of 1.2mm v's. orientation

In addition to measuring the total toughness of the fibres in pulling out along their full embedded lengths, a measure of toughness up to a certain value of crack-width is probably more relevant to the proposed model. Average values of post-fracture toughness up to a crack width of 1.2mm against fibre orientation are shown in Figure 5. It appears that as the fibre orientation with respect to loading is increased, the corresponding energy absorbed up to a certain crack width increases to a maximum (approximately 10-20 degrees) and then reduces. In addition, increasing the fibre embedded length also significantly increases toughness values at these orientations. However. at larger orientations toughness values appears to be converging towards a constant value irrespective of fibre embedded length.

Comparison of the results of Figure 4 and 5 may have important implications for the flexural response and toughness characterisation of SFRC. At small crack-widths (i.e. the area of a flexural load/deflection response up to I30) fibre orientation has a significant effect on toughness. However, as the crack-width increases this effect diminishes, so that at larger crack-widths (>7mm) there is no significant orientation effect. This result is suggesting that up to I30 the combination of fibre embedded length, and end-deformation, as well as fibre orientation are significant parameters for optimising post-fracture performance. However, for larger toughness indices the effect of fibre orientation diminishes and toughness performance is only influenced by fibre end deformation and embedded length.

5 Conclusions

A single fibre pull-out test has been developed in which the fibre is embedded within an uncracked matrix. The test allows the complete pull-out load/crack-width response to be recorded. Initial results suggest that this test provides a more realistic fibre/matrix stress distribution, compared to previous pull-out tests, for use in modelling flexural behaviour of a fibre reinforced sprayed concrete.

Whilst fibre orientations cannot be controlled (with the exception that the spraying process orientates fibres in a partially 2D array) the effect on the pull-out force, and hence toughness, depends on the serviceability level required in terms of crack-width and/or deflection.

6 References

1. Mangat,P.S. and Gurusamy,K. (1987) Flexural strength of steel fibre-reinforced cement composites. *Journal of Material Science*, Vol. 22, pp. 3103-3110.
2. Wecharatana,M. and Shah,S.P. (1983) A model for predicting fracture resistance of fiber reinforced concrete. *Cement and Concrete Research*, Vol. 13. pp. 819-829.
3. Hillerborg,A. (1980) Analysis of fracture by means of the fictitious crack model, particularly for fibre reinforced concrete. *International Journal of Cement Composites*, Vol. 2, pp. 177-184.
4. Krenchel,H. (1978) Fibre spacing and specific fibre surface, in *Testing and Test Methods for Fibre Cement Composites*, (ed. R.N. Swarmy), The Construction Press, London, pp. 69-79.
5. Banthia,N. and Trottier,J.F. (1995) Concrete reinforced with deformed steel fibers part II: toughness characterisation. *ACI Materials Journal*, Vol. 92, pp. 146-154.
6. EFNARC. (1993) Specification for sprayed concrete, Final Draft. EFNARC, Hampshire, UK.

PART FOUR
MATERIALS

13 DRY SHOTCRETE WITH RAPID HARDENING CEMENT AND HUMID AGGREGATES

Th. Eisenhut and H. Budelmann
Fachgebiet Baustoffkunde, Univ.-Gh Kassel, Germany

Abstract
The report deals with the properties of a new spray cement, the qualification tests of the processing of the cement with humid aggregates and the maintenance of the quay of the Eder sea in Hessen, Germany.
 For the shotcrete, the new spray cement CHRONOLITH®S has been used. A special ability of this cement is that there are no accelerators needed at all. To reduce the costs of the maintenance, humid aggregates were used instead of dry ones. The combination of short setting time of the cement and humid aggregates forced processing times less than 1:45 min after dosing. Therefore, a programmable dosing, weighing and mixing installation has been developed which guarantees processing times including hoisting time of the mixed materials to the spray nozzle of 90 sec. This preventive measures allows continuous loading and transferring of shotcrete. The results of the qualification and soundness tests turn out that the strength development of the shotcrete is in the range of J_2 of the ''Richtlinie Spritzbeton des Österreichischen Betonvereins''. After 24 h the compressive strength of the shotcrete was 20 N/mm² and after 28 d, 38 N/mm².
Keywords: Dry shotcrete, spray cement, humid aggregates, rapid hardening cement, programmable dosing, weighing and mixing installation.

1 Introduction

In the past, dry shotcrete was produced with standard cements and various accelerators to get rapid and early strengths. The use of accelerators may lead to a reduction of ultimate compressive strength, damage of health and endangering environment, whereas the use of humid aggregates reduces costs and the occurence of dust compared to dried ones. This choice of the humidity of aggregates influences the choice of the machinery. The machinability of humid aggregates and cement depends on the setting time of the cement and the dosing, weighing and mixing of the installation. These requirements lead to a new spray cement and a programmable dosing, weighing and mixing installation.

2 Spray cement

There are several possibilities to produce binders that fulfil the requirements for shotcrete [1, 2, 3]:
- Grounded pure cement with additives
- Grounded pure cement clinker phases
- Grounded pure cement clinker phases with additives
- Grounded special cement clinker phases
- Grounded special cement clinker phases with additives.

The used additives should influence the properties of the shotcrete that cannot be influenced by pure cement clinker phases, for example, to regulate the setting time.

By using pure clinker phases as binders for shotcrete the setting regulator gypsum/anhydrite can be left out. This measure influences the setting time of the dry shotcrete mixture after the contact with water. Furthermore, while fabricating a cement by means of a suitable selection of starting materials and a special burning process, it is possible to get a binder that fulfils the requirements for shotcrete.

In general, it is not possible to produce a shotcrete binder by using only cement clinker phases to get all properties of the shotcrete as:
- Low rebound
- Tolerance for change of the water/cement content
- No increase of alkalinity
- Low endangering (health of operating personnel and industrial hygiene)
- Compressive strength at the age of 28 d like ordinary concrete
- No elutriation and consequently high durability.

Therefore, additives are necessary. Usually these additives are accelerating admixtures which reduce rebound and dust. In addition stabilisators e.g. products on the basis of amorphous silica [4] can be used to prevent the shotcrete from bleeding, separating and, above all, to increase the binding property and at the same time reduce rebound and dust.

Accelerators regulate the setting behaviour of the cement by contact with water. The accelerators commonly used are on the basis of [4, 5]:
- Silicate
- Aluminate
- Carbonate (cement clinker phases)
- Inorganic neutral salts
- Organic neutral salts
- Sulphate reduced cement

The silicate, aluminate and carbonate based accelerators reduce the compressive strength of the shotcrete at the age of 28 d compared with cement without accelerators. The silicate based accelerators lead to a further formation of calcium silicates. The aluminate based accelerators, carbonate based and inorganic neutral salts react with the cement to a formation of ettringite.

The organic neutral salts also influence the formation of ettringite and the solubility of the calcium hydroxide. This accelerator also reduces the compressive strength of the shotcrete.

Due to the lack of sulphate, in gypsum reduced cement the C_3A will react rapidly with water by formation of calcium aluminate hydrates without a reduction of compressive strength. This may lead to other problems which will not be discussed here in detail.

The mineralogical composition of the new spray cement CHRONOLITH®S is based on grounded special cement clinker phases with additives. The additives have set accelerating functions. On the one hand, the problem was the combination of the additives and the special cement clinker phases, and on the other hand, the tolerance towards humid aggregates in combination with the set accelerating time.

The spray cement CHRONOLITH®S has a low C_3A-content, and as a consequence of this, it has a high sulphate resistance. This cement is produced on the base of portland cement clinker phases. The amount of portland cement clinker phases is ≥ 95 % [6]. The setting time begins at the age of 1:45 min and ends 3:00 min after contact with water measured according to [7]. The compressive strength of the cement with a water/cement ratio of 0.4 is shown in table 1 [7, 8, 9].

Table 1. Compressive strength of mortar with the cement CHRONOLITH®S

Age	Compressive strength (N/mm²) measured according to [7]
1 h	0.7
3 h	1.0
6 h	2.4
1 d	20
7 d	38
28 d	49

Due to the fact that no accelerators are needed, the load of the ventilation air and the ecological damage can be reduced. Furthermore, the elutriation of calcium, potassium, sodium and aluminate of the CHRONOLITH®S is reduced compared to CEM I 32.5 R with a sodium-aluminium accelerator. The elutriation reduction of calcium is about 63 %, of potassium 72 %, of sodium 48 % and aluminium 16 % [8, 10].

3 System-requirements for preparation and conveyance of humid aggregates

The choice of using the spraying method has created controversial discussions. The use of dry or wet shotcrete depends on economical and technical facts.

By using the wet method, there is a reduction of dust development and rebound if optimised equipment is used compared to the dry method. On the ohter hand the rebound is more influenced by the conductor at the spray nozzle, the composition of the starting/input material and the admission of air at the nozzle than by the spraying method [12]. Moreover, there is a high flexibility of the dry spraying method with longer processing time and without addition of retarders.

An important economical advantage of the dry spraying method is the possibility to combine the use of the new spray cement with humid aggregates ($w_g \approx 5$ wt.-%). The use of humid aggregates in combination with the dry shotcrete method reduces the dust development at the spray nozzle and in the loading and mixing area. As a consequence, the processing technique must be adapted to the setting time.

This consideration led to the development of a programmable dosing, weighing and mixing unit (see fig. 1). There is a twin chamber silo that can be charged separately with cement and humid aggregates. The forced mixing installation is set below the twin chamber silo. There, cement and aggregates are charged with separate trigger screws and the whole mixing installation is weighed. Furthermore the quantities of cement, aggregates and water, if needed, are programmable. An output of 0.5 to 7 m³/h can be obtained.

The unit has a mixing time of 45 s and a maximum hoisting time of 45 s. A total processing time of 90 s is guaranteed. This time is sufficiently below the start setting time of 1:45 min.

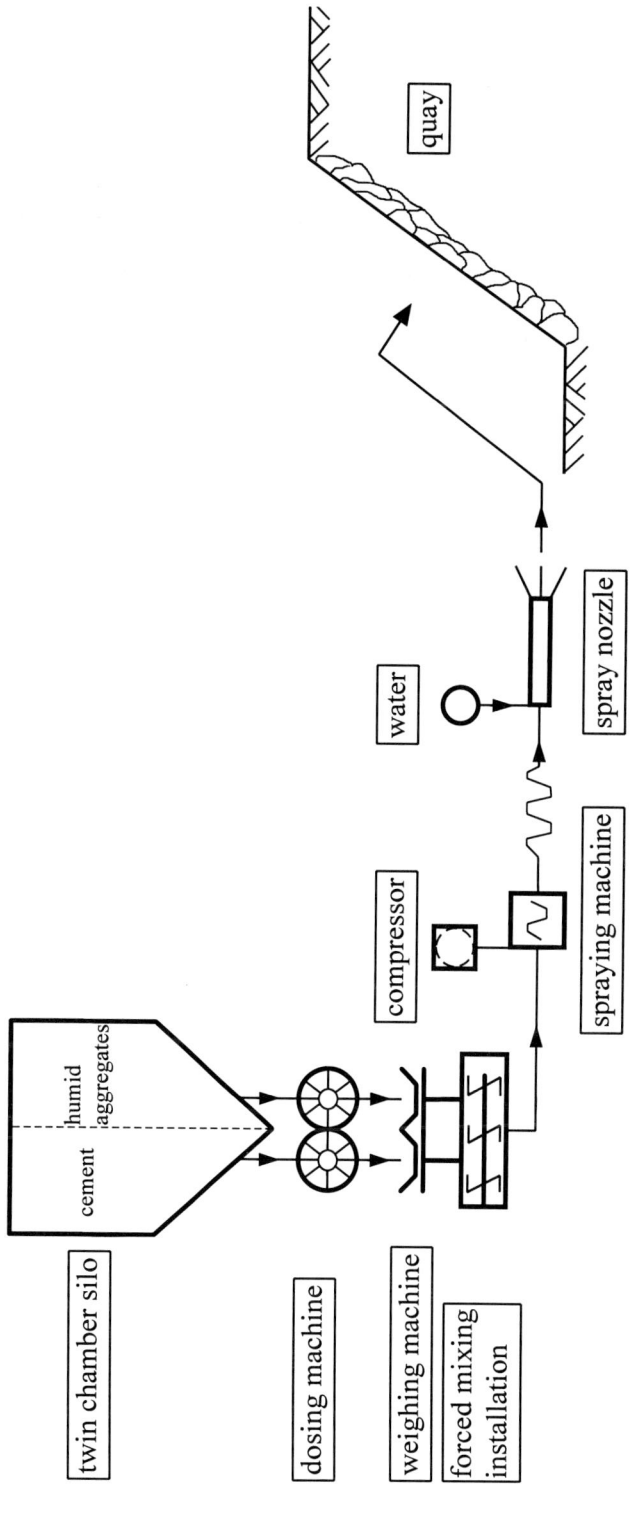

Fig. 1. Programmable mixing and shooting installation

4 Experimental investigations

4.1 Experimental program

The testing program includes the measurement of the shotcrete and ambient temperature for the first 24 h and the amount of rebound. The compressive strengths of the shotcrete has been determined after 2 min, 6 min, 15 min, 30 min, 1 h, 2 h, 3 h with the penetration needle, after 6 h with the HILTI setter method and after 1 d, 7 d and 28 d with cores according to [13].

4.2 Testing technique

A spray testing site has been builded of reinforced concrete plates. The horizontally and vertically fitted plates were fixed on a scaffolding and subdivided by wooden sheathings into 2 x 2 m (l x b) testing areas with a depth of 0.20 m [14].

Each test section was equipped with two temperature sensors [14], installed 7 and 14 cm below the expected surface of the shotcrete. An additional sensor measures the ambient temperature. The rebound was collected by using plastic sheet.

The development of the early compressive strength was determined with the penetration needle method and the HILTI setter method [13]. Other compressive strengths were determined from drilled cores (d = 100 mm, l/d = 1) taken from the plane shotcrete as well as from the spraying boxes [15].

With the penetration needle method, the resistance to penetrate of a standard needle is measured for depth of 1.5 cm. With this method it is possible to measure compressive strengths below 1.5 N/mm². In order to obtain a valid test, ten measurements are necessary to assess the compressive strength [13].

With the HILTI setter method, a pre-shoot screw bolt is pulled out from the shotcrete. The compressive strength is the relationship of the penetration depth to the extracting force. This method is suited to register compressive strengths from 2 to 15 N/mm². As for the penetration needle test, ten measurements are also needed [13].

Shotcrete with special requirements according to early load (6 min to 24 h) [13] can be separated into the range of J_1, J_2 and J_3. In the event of rapid occurence of active rock pressure, it sets high requirements on the development of compressive strength. In this case, the range of J_2 could be applied according to [13].

4.3 Test results

In order to receive comparable testing results, a basic experiment was performed with dry aggregates (w_g = 0 wt.-%).

The dry shotcrete mixture was 380 kg of cement and 1720 kg of aggregates (0/8 gradation according to [16]). The water content, which was not weighed, was about 200 kg. All testing areas had been moistened before spraying.

With this mixture of the fresh concrete, a rebound of 20.3 % of weight was measured.

After 24 h, the shotcrete temperature had dropped down to the ambient temperature. The maximum difference between shotcrete and ambient temperature was 7.3 K.

The compressive strength development until an age of the shotcrete of 3 h was determined with the penetration needle method, until 6 h with the HILTI setter method and beginning from an age of 12 h with drill cores.

In Fig. 2 the development of the temperature difference between the shotcrete, the ambient air and the development of the compressive strength are shown. At the beginning of the shotcrete hardening, both, the temperature difference of the shotcrete to ambient air and the compressive strength increased. Afterwards, the temperature decreased and after crossing the temperature difference maximum of 7.3 K the compressive strength stagnated at 0.67 N/mm². This stagnation lasts to about 6 h.

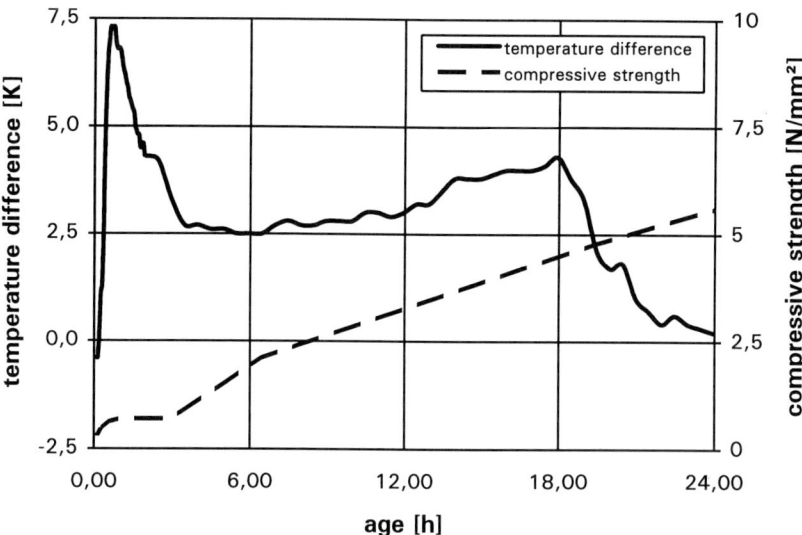

Fig. 2. Development of the temperature difference and of the compressive strength of the shotcrete

Compared with results obtained with dry aggregates, the use of humid aggregates ($w_g \approx 5$ wt.-%) led to an increase of the compressive strength (see Fig. 3). There was also a resting phase (30 min until 6 h) in the development of compressive strength with humid aggregates, so that the compressive strength of the shotcrete dropped down in the range of J_1 [13] during the resting phase. But afterwards the strength development was in the range of J_2 again. The requirement was to be in the range of J_2 at least.

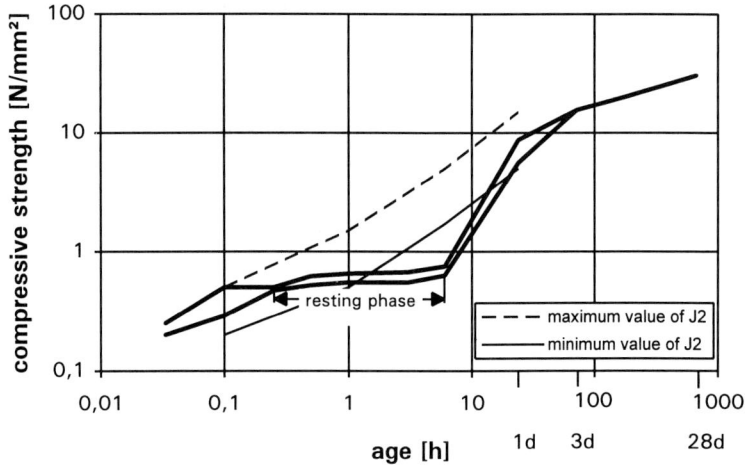

Fig. 3. Results of compressive strength

5 Practical application at site

The described technology was used to the quay of the Eder storage pond, a 18.00 m high sloped dry wall. The walls bulged out and threatened to cave in. The wall was in need of repair to restore the stability. The shotcrete was used to safeguard the quay. No accelerators should be used due to water's protection reason. The wall was excavated in small sections. The loose ground behind the wall was covered by a 25 cm to 30 cm thick protective layer of shotcrete and coated again with stones. The whole procedure consisted of three partial sectors with a total of 140 m of the quay to be repaired.

The supervising governmental agency demanded a compressive strength of 25 N/mm² at a shotcrete age of 7 d and 35 N/mm² at an age of 28 d. In addition, a development of the compressive strength in the range of J_2 was needed [13].

The compressive strength was determined only at spraying boxes [15]. The humidity of the aggregates varied between 4.0 and 5.3 percentage by weight. The density of the drill cores varied in a range of 2.24 t/m³ to 2.31 t/m³. The resting phase in the development of compressive strength observed in the experiments before, was displaced to 15 min to 3 h instead of 30 min to 6 h (see Fig. 4). The compressive strength at an age of 7 d was determined in a range of 30 N/mm² to 40 N/mm², and at an age of 28 d in a range of 37 N/mm² to 41 N/mm².

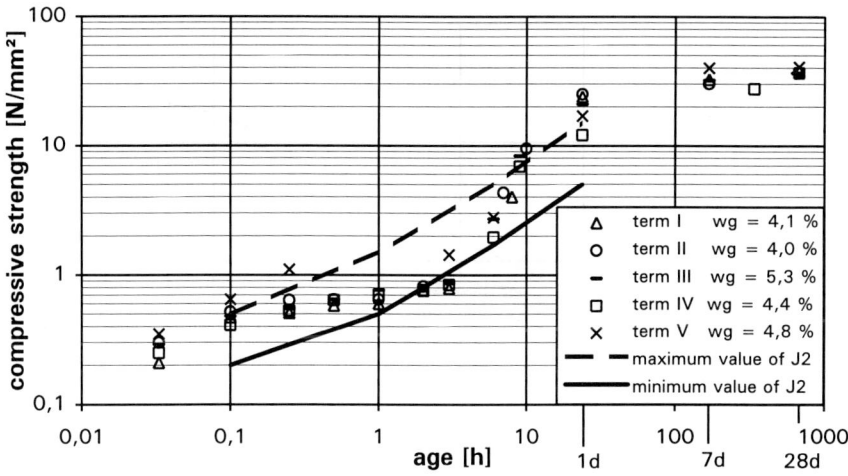

Fig. 4. Development of compressive strength at the Eder sea

6 Summary

As shown, it is possible to produce shotcrete with humid aggregates and the new spray cement that does not contain accelerators. The development of compressive strength was in the range of J_2 according to [13]. The compressive strength at an age of 28 d is 38 N/mm² in the range of a concrete B 25 according to DIN 1045.

Finally, the combination of short setting time of the spray cement, humid aggregates and the programmable dosing, weighing and mixing installation is an efficient and economical way of producing shotcrete.

7 Reverences

1. Löschnig, P.; Müller, L.; Unsin, J. (11.1993) *Entwicklung eines umweltgerechten Spritzbetonsystems*. Firmenprospekt, Leimen.
2. Maidl, B.; Berger, Th. (1995) Empfehlungen für den Spritzbetoneinsatz im Tunnelbau. *Bauingenieur*, Nr. 70, S. 11 - 19.
3. Schmidt, M.; Felten, V. (1993) Europäisches Forschungsvorhaben zur Entwicklung eines umweltgerechten Spritzbetonsystems, in *Berichtsband der 4. internationalen Fachtagung Spritzbetontechnologie*. Institut für Baustofflehre und Materialprüfung, Universität Innsbruck.
4. Mai, D. (1993) Zusatzmittel für die moderne Spritzbetontechnologie. in *Berichtsband der 4. internationalen Fachtagung Spritzbetontechnologie*, Institut für Baustofflehre und Materialprüfung, Universität Innsbruck.
5. Reul, H. (1991) *Handbuch Bauchemie*. Verlag für chem. Industrie, Augsburg.

6. Deutsches Institut für Bautechnik (1995) Allgemeine bauaufsichtliche Zulassung, *Zulassungs-Nr. Z-3.12-143*. Berlin, DIBt.
7. Europäisches Komitee für Normung. (03.1990) *Prüfverfahren für Zement. Teil 1: Bestimmung der Festigkeit, Teil 3: Bestimmung der Erstarrungszeiten und Raumbeständigkeit*. Brüssel, DIN EN 196: Teil 1, 3.
8. Heidelberger Zement AG (1995) Umweltverträglicher Spritzzement für das Trockenspritzverfahren. *Technisches Merkblatt für Sorten-Nr. 01152*.
9. Deutsches Institut für Normung. (10.1994) *Zement. Teil 1: Zusammensetzung, Anforderungen*. Berlin, DIN 1164: Teil 1.
10. Österreichisches Normungsinstitut. (12.1990) *Eluatklassen (Gefährdungspotential) von Abfällen*. Wien, ÖNORM S 2072.
11. Eichler, K. (1994) Umweltfreundlicher Spritzbeton. *Tunnel*, Heft 1, S. 33 - 37.
12. Ruffert, G. (1987) Entwicklungstendenzen in der Spritzbetontechnik. *Beton- und Stahlbeton*, Heft 7, S. 186 -190.
13. Österreichischer Betonverein. (01.1989/06.1991) *Richtlinie „Spritzbeton". Teil 1: Anwendung, Teil 2: Prüfverfahren*. Wien.
14. Eisenhut, Th.; Budelmann, H. (03.1995) *Trockenspritzbeton mit eigenfeuchtem Zuschlag ohne Beschleunigerzugabe*. Ergebnisbericht, FG Baustoffkunde, Kassel.
15. Deutsches Institut für Normung. (03.1992) *Spritzbeton. Herstellung und Güteüberwachung*, Berlin, DIN 18551.
16. Deutsches Institut für Normung. (07.1988) *Beton und Stahlbeton. Bemessung und Ausführung*, Berlin, DIN 1045.

14 POLYOLEFIN FIBRE REINFORCED WET-MIX SHOTCRETE

D.R. Morgan
AGRA Earth and Environmental Ltd, Vancouver, Canada
L.D. Rich
3M Company, Minnesota, USA

Abstract
Polymer fibres (mainly fibrillated polypropylene) have been used for nearly a decade in wet-mix shotcretes at addition rates varying between 0.10 and 0.66 percent by volume, i.e. 0.9 to 6 kg/m^3. A new non-fibrillated polyolefin fibre has been developed which can be added to concrete and wet-mix shotcrete mixtures at considerably higher addition rates. This has been made possible because of the characteristics of the fibre (lower aspect ratio and lower specific surface area compared to fibrillated polypropylene fibres) plus a unique fibre dispensing system. A study was undertaken to assess possible addition rates of the new polyolefin fibre to wet-mix shotcretes and to evaluate the plastic and hardened properties of such shotcretes compared to a plain control shotcrete mixture with no fibre reinforcement. In this study the polyolefin fibre was added to the shotcrete at rates of 1.0 and 2.0 percent by volume (9.1 and 18.2 kg/m^3). Test panels were shot from which specimens were procured for determination of compressive and flexural strength, toughness, boiled absorption and volume of permeable voids. Test results showed that good residual load carrying capacity after first crack in flexural toughness testing (toughness performance level) could be obtained at 1.0 and 2.0 percent by volume polyolefin fibre addition rates. Limitations on pumpability of the mixture with 2.0 percent by volume of polyolefin fibre however indicated that in the field fibre addition rates would likely be limited to about 1.25 to 1.5 percent by volume for the type of shotcrete equipment commonly being used in North America. Such mixtures are attractive for use in a variety of applications where toughness and control of cracking are important.
Keywords: Fibre, polyolefin, shotcrete, toughness.

1 Introduction

Since the early 1970's, steel fibre reinforced shotcrete has been used in lieu of wire mesh reinforced shotcrete in a variety of applications, such as ground support in tunnels and mines, slope stabilization, lining canals and other water-retaining structures, seismic retrofit, and infrastructure rehabilitation[1].

Steel fibre has generally been used in wet-mix shotcretes at addition rates in the 0.5 to 1.0 percent by volume range (39.3 to 78.6 kg/m^3) with 0.76 percent by volume (60 kg/m^3) addition rates being most common. In the 1980's, some polymer fibres (mainly monofilament and fibrillated polypropylene or nylon fibres) started to find use in wet-mix shotcretes at low addition rates, e.g. 0.1 percent by volume (0.9 kg/m^3) for control of plastic shrinkage cracking and to enhance *green strength* of freshly applied shotcrete. At such low fibre addition rates the effects of the fibre on residual flexural strength after first crack are very small, i.e. the toughness performance level [2] was not much better than that in plain (non-fibre reinforced) shotcrete.

In the mid 1980's studies were conducted on polymer fibre reinforced wet-mix shotcretes with fibrillated polypropylene fibres added at rates of between 0.44 and 0.66 percent by volume (4 and 6 kg/m^3) [3,4]. These studies demonstrated that at these higher polymer fibre addition rates residual flexural strengths (toughness performance levels) could be obtained which were equivalent in performance to certain lesser performing types of steel fibre reinforcement at equivalent percent by volume addition rates. Toughness performance levels were, however, inferior compared to the better performing types of steel fibre reinforced shotcretes. It was apparent that higher addition rates of polymer fibre reinforcement would be required if toughness performance levels approaching those of the better performing steel fibre reinforced shotcretes were to be achieved. Practical experience in the field, however, demonstrated that addition rates of more than 0.66 percent volume (6 kg/m^3) were difficult to achieve with fibrillated polypropylene fibres; most contractors were reluctant to use such fibres at more than about 0.44 or 0.55 percent volume (4 or 5 kg/m^3) because of difficulties in consistently batching, mixing, pumping and shooting such mixtures without blockages in the pump and hose.

It became apparent that a polymer fibre with a lower aspect ratio (length to equivalent diameter) and specific surface area and/or dispensing system would be required if higher addition rates of polymer fibres were to be achieved. A new monofilament polyolefin fibre, with a unique, patented dispensing system had been successfully developed for use in concrete at addition rates of up to 2.0 percent by volume (18.2 kg/m^3). It was decided to evaluate the suitability of this fibre for use in wet-mix shotcrete. The fibre was added at site to a transit mixer at addition rates of 1.0 and 2.0 percent by volume (9.1 and 18.2 kg/m^3). Test panels were shot and test specimens procured for determination of properties of the plastic (fresh) and hardened shotcrete.

2 Mixture proportions

Three shotcrete mixtures were selected for comparative evaluation; a plain control shotcrete mixture with no fibre; a mixture with 1.0 percent by volume (9.1 kg/m^3) polyolefin fibre and a mixture with 2.0 percent by volume addition of polyolefin fibre. The fibre used was a monofilament fibre, 25 mm long 0.38 mm diameter (i.e. aspect ratio of 66). The fibre was supplied in 55 mm diameter bundles, 25 mm high, wrapped in a *timed-release* water dispersable tape (colloquially referred as *pucks*). The fibre was supplied in a 9.1 kg box containing about 240 of these *pucks*. Figure 1 shows a typical fibre *puck*.

Fig. 1. Polyolefin fibre *puck* containing 25 mm long 0.38 mm diameter fibres.

A shotcrete mixture design, similar to that commonly used in Canada for steel fibre reinforced wet-mix shotcretes, was used. The shotcrete mixture designs used are given in Table 1. The plain control mixture was designated as Mix A. The mixtures with 1.0 and 2.0 percent by volume polyolefin fibre addition were designated as Mix B and Mix C respectively.

3 Shotcrete supply and application

Wet-mix shotcrete was batched by a commercial transit mix supply company located about 20 minutes haul time from the test site. All the shotcrete ingredients with the exception of the polyolefin fibre and the superplasticizer, were transit mix batched at the plant. The properties of the plastic shotcrete mixtures before and after fibre addition are given in Table 2.

The shotcrete was supplied in 2.0 m^3 loads. The plain shotcrete was reversed in the transit mixer nearly to the point of discharge. The fibre *pucks* were dumped into the mixer, onto the shotcrete, and the load mixed for approximately seven minutes at full mixing speed. The fibre *pucks* mixed through the load (almost like pieces of large coarse aggregate). After 3 to 4 minutes the timed release water dispersable tape broke down, releasing the now distributed fibres into the shotcrete. Within about 7 to 8 minutes from the time of introduction of the fibres into the transit mixer, the fibres were well dispersed throughout the load. No incompletely dispersed fibres were seen

Table 1. Shotcrete mixture proportions

Material	Batch proportions (kg/m^3) based on SSD aggregates		
Mix designation	A	B	C
Mix description	Plain	1.0% vol. fibre	2.0% vol. fibre
Portland cement, type 10	400	400	400
Silica fume (uncompacted)	48	48	48
Coarse aggregate (SSD) 10 x 2.5 mm	480	480	480
Concrete sand	1110	1110	1110
Water	190	190	200
Water reducing admixture Pozzolith 122R	1760 mL	1760 mL	1760 mL
Superplasticizer Rheobuild 1000	0 mL	0 mL	1000 mL
3M polyolefin fibre 25 mm x 0.38 mm dia.	0	9.1	18.2
Air entraining admixture, as required for air content: as batched: as shot:	10±1% 4±1%	10±1% 4±1%	10±1% 4±1%
Total mass	2230	2239	2258

Table 2. Shotcrete plastic properties

Mix No.	Mix description	Temperature °C		Slump mm		Air content %		
		Ambient	Shotcrete	Before fibre addition	After fibre addition	Before fibre addition	After fibre addition	As shot
A	plain shotcrete	1	10	100	-	11.0	-	4.0
B	1.0% fibre volume	1	10	100	25	11.0	9.5	5.6
C	2.0% fibre volume	1	10	90	75*	9.0	8.0*	4.0

* After fibre, superplasticizer and 10 L/m^3 water addition.

during discharge of Mix B with 1.0 percent by volume fibres; some incompletely dispersed fibre bundles were, however, seen in Mix C with 2.0 percent by volume fibres and this mix was given another 10 L/m^3 of water addition plus 1 L of superplasticizer and additional mixing time.

The shotcrete was intentionally supplied at an air content of 10±1% in order to take advantage of the following characteristics of high air content in as batched shotcrete mixtures, as demonstrated by Beaupré et al [5, 6]:
- a high paste volume (paste being comprised of cement, silica fume, water, chemical admixtures and air) to facilitate coating of the fibres;
- a workability enhancing and pumping aid (high air contents increase slump and reduce required pump pressures);
- compaction during shooting; i.e. as the shotcrete impacts on the receiving surfaces the air content is reduced to 3 to 5 % and the slump is markedly reduced.

Actual slumps and air contents for the different mixtures, before and after shooting are shown in Table 2. Both the plain shotcrete, Mix A and Mix B with 1.0 percent by volume addition shot readily with no pump blockages. The nozzleman was able to shoot two standard test panels with Mix C with 2.0 percent by volume fibres, but some problems with fibres "matting" and plugging at the reducing line from the pump hopper to the shotcrete hose were encountered during shooting of the remainder of this 2 m^3 load onto a rock slope. It was apparent that 2.0 percent by volume polyolefin fibre addition, while frequently used in cast concrete construction, was likely an excessive fibre addition rate for the type of equipment used. The pump used was a Schwing 750 RD swing-tube pump, with a 50 mm internal diameter shotcrete hose. This is a commonly used type of equipment for wet-mix shotcrete application by hand nozzling in North America.

4 Hardened shotcrete properties

Two standard 600 x 600 x 125 mm test panels were shot for each of the three mixtures. The test panels were stored in the field under curing blankets with space heaters for 3 days before being transported (in their forms) to the laboratory for moist curing at 23°C and 98±2% R. H. until the time of testing. Specimens were diamond sawed or cored from the test panels for testing at age 7 days (unless otherwise indicated) for determination of:
- compressive strength (7 and 28 days)to ASTM C 39;
- boiled absorption and volume of permeable voids to ASTM C 642;
- flexural strength and toughness to ASTM C 1018 [7], Japanese toughness and toughness factor [8] and toughness performance level [2];

Results of the compressive strength, boiled absorption and volume of permeable voids tests are given in Table 3.

Table 3 Compressive strength and ASTM C642 test results

Mix No.	Mix description	Compressive strength at 7 days	Compressive strength at 28 days	ASTM C 642		
		MPa	MPa	Absorption after immersion and boiling, %	Bulk S.G. after immersion and boiling	Volume of permeable voids, %
A	plain shotcrete	35.9	52.8	4.6	2.262	10.0
		38.7	47.3	5.0	2.254	10.8
	average	37.3	50.0	4.8	2.258	10.4
B	1.0% fibre vol.	36.5	41.0	5.0	2.222	10.5
		32.3	41.5	4.6	2.245	9.8
	average	34.4	41.3	4.8	2.233	10.2
C	2.0% fibre vol.	28.5	41.0	6.8	2.188	14.0
		27.1	40.5	6.5	2.197	13.4
	average	27.8	40.8	6.7	2.193	13.7

The compressive strength and ASTM C 642 tests were conducted on 75 mm diameter x about 100 mm cores extracted from the test panels. Core strengths were corrected to equivalent 2:1 length:diameter cores using the ASTM C 42-94 strength correction factors.

Results of flexural strength and toughness tests are given Tables 4 and 5. These tests were conducted on 100 x 100 x 350 mm beams cut from the test panels. The tests were conducted in accordance with the ASTM C 1018-94b test procedures [7]. Load vs. deflection curves for Mix B and Mix C are shown in Fig. 2 and Fig. 3 respectively. Table 4 presents the results analyzed in accordance with the ASTM C 1018-94b procedures for calculation of toughness indices and residual strength factors [7]. Table 5 presents the analysis of the results (from the same load deflection curves) in accordance with the JSCE-SF4 procedure [8] and Toughness Performance Levels calculated in accordance with the method proposed by Morgan et al [2]. The data in Tables 4 and 5 provides a comparison of these different methods currently being used in North America for characterising toughness of fibre reinforced shotcretes.

5 Discussion of Test Results

To obtain an appreciation of the significance of the test results, the reader is referred to Table 6, which provides a summary of performance requirements commonly specified the in the 1980's and 1990's in Canada for fibre reinforced shotcrete for ground support applications. [10].

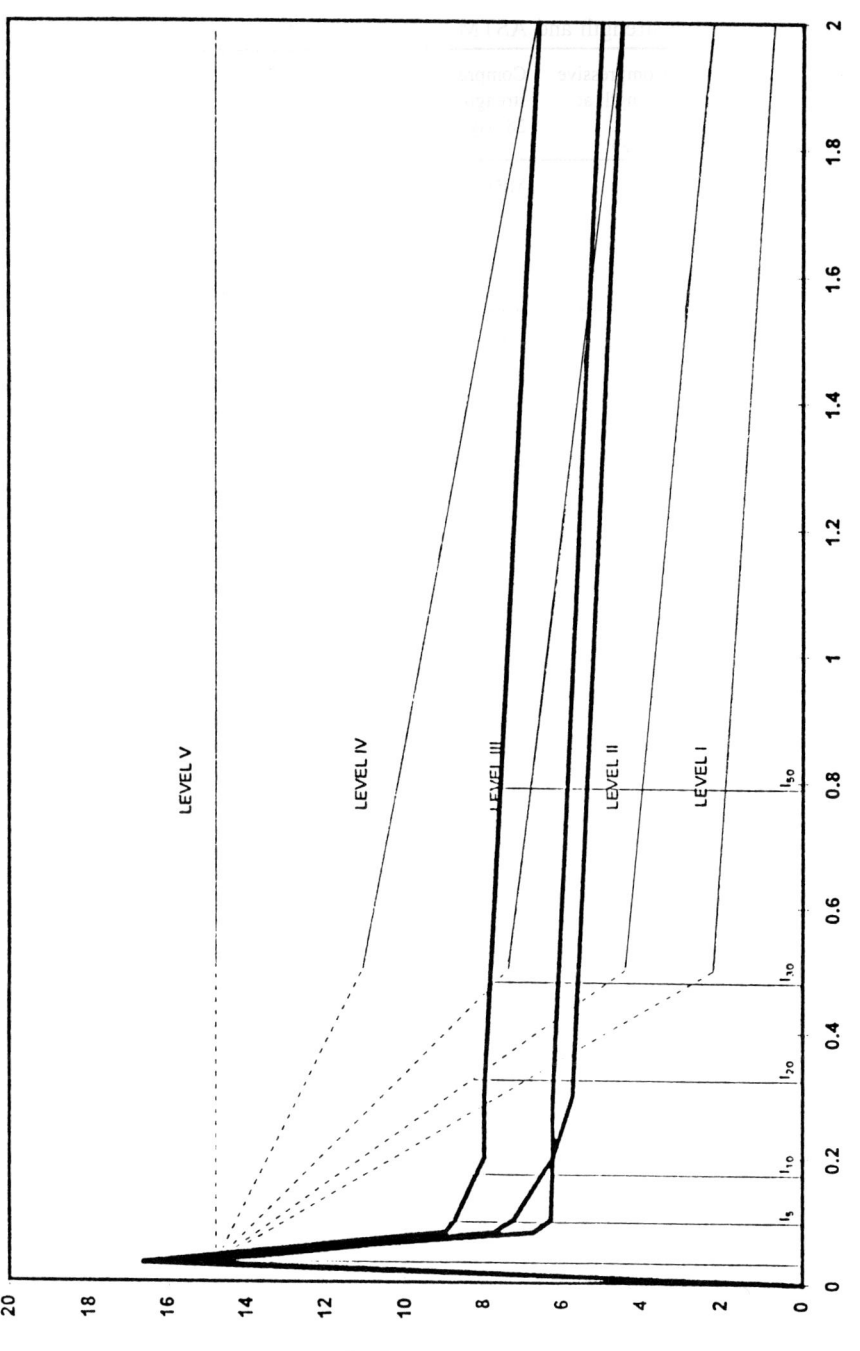

Fig. 2. Load vs. deflection curves for mix B with 1.0 percent by volume polyolefin fibre.

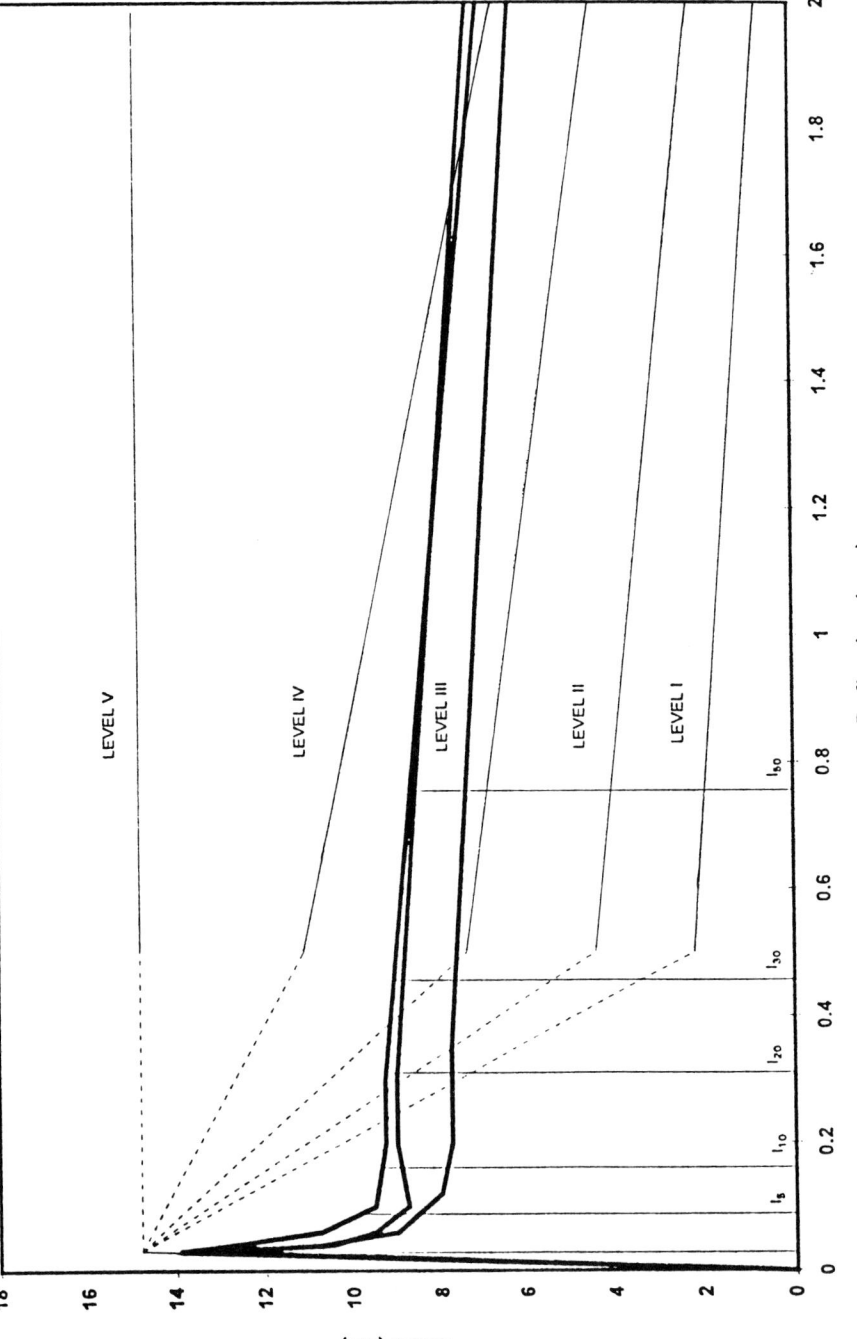

Fig. 3. Load vs. deflection curves for mix C with 2.0 percent by volume polyolefin fibre.

Table 4. Flexural strength and flexural toughness to ASTM C 1018-94b

Sample No.	Mix description	7 day flexural strength (MPa)	ASTM C 1018 toughness indices					ASTM C 1018 residual strength factors			
			I_5	I_{10}	I_{20}	I_{30}	I_{50}	$R_{5,10}$	$R_{10,20}$	$R_{20,30}$	$R_{30,50}$
A1	plain	4.02									
A2	shotcrete	4.51									
A3		4.98									
Avg.		4.50									
B1	1.0% vol.	4.23	4.1	6.1	10.2	14.1	21.6	40.0	41.0	39.0	37.5
B2	fibre	4.50	4.2	6.9	11.8	16.6	25.9	54.0	49.0	48.0	46.5
B3		4.36	4.0	6.4	10.5	14.0	20.6	48.0	41.0	35.0	33.0
Avg.		4.36	4.1	6.5	10.8	14.9	22.7	48.0	43.0	41.0	39.0
C1	2.0% vol.	3.87	4.4	7.9	14.5	21.0	32.9	70.0	68.0	65.0	59.5
C2	fibre	3.70	4.3	7.3	13.0	18.6	29.5	60.0	57.0	56.0	54.5
C3		3.58	4.4	7.8	14.3	21.0	33.9	68.0	65.0	67.0	64.5
Avg.		3.70	4.4	7.7	13.9	20.2	32.1	66.0	62.0	63.0	59.5

Table 5. Flexural strength and flexural toughness to JSCE-SF4 and toughness performance level

Sample No.	Mix description	7 day flexural strength (MPa)	JSCE-SF4 [1]		Morgan et al [2]
			Toughness (kN/mm)	Toughness Factor (MPa)	Toughness Performance Level *
A1	Wet-mix, no	4.02			0
A2	fiber	4.51			0
A3		4.98			0
Avg.		4.50			0
B1	Wet-mix, 1%	4.23	11.6	1.57	II-III
B2	vol. fibre	4.50	14.8	2.00	III
B3		4.36	10.5	1.45	II
Avg.		4.36	12.3	1.67	II-III
C1	Wet-mix, 2%	3.81	16.0	2.19	III
C2	vol. fibre	3.70	14.2	1.93	III
C3		3.58	16.2	2.19	III
Avg.		3.70	15.5	2.10	III

* Note: Toughness performance level calculation based on a design flexural strength of 4 MPa at 7 days.

Table 6. Typical fibre reinforced shotcrete performance specifications

Property	Test method	Requirements
Compressive strength:	ASTM C-42	
1 day		10 MPa min
7 days		30 MPa min
28 days		40 MPa min
Flexural strength:	ASTM C-1018	
7 days		4 MPa min
28 days		6 MPa min
Toughness index	ASTM C-1018	
I_5		3.5 average
I_{10}		5.0 average
Boiled absorption	ASTM C-642	8% maximum
Volume of permeable voids	ASTM C-642	17% maximum

Compared against these performance requirements it can be seen that Mix A and Mix B satisfy the compressive strength requirements of 30 MPa at 7 days and 40 MPa at 28 days. Similarly Mix A and Mix B satisfy a 4 MPa at 7 days flexural strength requirement. Mix C, with 2.0 percent by volume fibre addition has a lower than specified 7 day compressive strength and flexural strength. This is attributed primarily to the addition of 10 L/m^3 of water to retemper Mix C (in conjunction with addition of 1 L/m^3 superplasticizer) to bring it to a consistency suitable for pumping.

Mixes A, B, and C all satisfied commonly specified maximum values of absorption after boiling of 8 percent and volume of permeable voids of 17 percent.

If the toughness indices I_5 and I_{10} for Mix B and Mix C given in Table 4 are compared against the limits given in Table 6, then it can be seen that the I_5 and the I_{10} limits are satisfied. It should, however, be noted that a proposal is currently before the ASTM C 1018 committee that the I_5 index (and residual strength factors calculated from it) be abandoned. Of more interest are residual strength factors at greater deformations, such as $R_{10,20}$, $R_{20,30}$ and $R_{30,50}$. The beneficial attributes of the polyolefin fibre are well demonstrated by these parameters as well as by the Japanese Toughness Factor and the Toughness Performance Levels given in Table 5.

An appreciation of the significance of achieving a Toughness Performance Level of II to III for Mix B, with 1.0 percent by volume of polyolefin fibre and Toughness Performance Level of III for Mix C, with 2.0 percent by volume of polyolefin fibre can be obtained by comparing the data in Table 5 with previously published data for a range of different types and addition rates of steel fibres in shotcretes [2]. In brief, at 1.0 percent by volume the polyolefin fibre provides a superior Toughness Performance Level to certain lesser performing types of steel fibres and approaches the performance of certain better performing types of steel fibres added at around 0.7 percent by volume. A 2.0 percent by volume addition rates the polyolefin fibre produced similar performance to certain better performing types of steel fibres added at around 0.7 percent by volume.

6 Conclusions

This study has demonstrated that:

The new non-fibrillated 25 mm long x 0.38 mm diameter polyolefin fibre evaluated can be added to a wet-mix shotcrete in a transit mixer and mixed at fibre addition rates of 1.0 and 2.0 percent by volume (9.1 and 18.2 kg/m^3).

The mixture with 1.0 percent by volume of polyolefin fibre displayed good pumping and shooting characteristics; some difficulty was, however, encountered in pumping and shooting the mixture with 2.0 percent by volume of polyolefin fibre, as it tended to block at the point where the reducing line from the pump was attached to the 50 mm internal diameter shotcrete hose. This contrasts with fibrillated polypropylene fibers which are difficult to pump and shoot at fiber addition rates above about 0.5 percent by volume (4.5 kg/m^3).

The shotcrete mixture with 1.0 percent by volume of polyolefin fibre satisfied commonly specified performance characteristics for fibre reinforced wet-mix shotcretes with respect to compressive and flexural strengths, boiled absorption and volume of permeable voids.

At 1.0 percent by volume (9.1 kg/m^3) fibre addition rates the polyolefin fibre is capable of producing good Toughness Performance Levels intermediate in performance between certain *lesser* and *better* performing types of steel fibres added at rates of around 0.7 percent by volume (55 kg/m^3).

Based on this study the polyolefin fibre was subsequently successfully used in two large-scale field prototype studies, where it was added at fibre addition rates between 1.0 and 1.5 percent by volume. A fibre addition rate of about 1.25 percent by volume was indicated as being about optimum for this particular fibre for the types of shotcrete mixtures and placing equipment commonly being used for hand nozzling applications in North America. The results of these field prototype studies will be presented in a separate publication.

7 Acknowledgements

This study was sponsored by 3M Construction Markets Division, St. Paul, Minnesota, USA and was conducted by AGRA Earth & Environmental Limited, Vancouver, British Columbia, Canada. The polyolefin fibre used was supplied by the 3M Company.

15 SPRAYED FIBRE CONCRETE: THE WAY FORWARD IN CIVIL ENGINEERING

M.J. Scott
Geotechnics, Ove Arup & Partners, Cardiff, UK

Abstract
"Sprayed Concrete" is a mixture of cement, aggregate and water projected at high velocity from a nozzle into place to produce a dense homogeneous mass. This form of construction was first used in the early part of this century. It is usually reinforced with mesh.

A fairly recent development has been to replace the reinforcement mesh with fibres, usually steel but fibreglass and polypropylene have also been used. Despite these advantages the Civil Engineering industry has been slow to adopt this technique. This is surprising given the advantages of Sprayed Fibre Concrete (SFC) over traditional mesh reinforced sprayed concrete, ie. faster construction, better stress and strain properties and higher flexural strengths.

Research by the author has revealed that no British Standard, design guides or specifications exist for SFC. In fact, only the Japanese have so far produced a specification for this material, Reference 1.

This paper presents two recent case studies where the use of SFC applied by the dry mix process has enabled two major Civil Engineering projects to be completed with an aesthetic, cost effective solution. The advantages of SFC over more conventional mesh reinforced sprayed concrete solutions is discussed.

Keywords: fibres, mines, rock cuttings, sprayed concrete, tunnels.

1 Introduction

1.1   Case study 1 : Rhuallt Hill
Rhuallt Hill was one of the last sections of road to be upgraded to dual carriageway standard on the A55 North Wales highway. The scheme included two rock cuttings up to 27m deep with up to 1km length of rock face. Side slopes were designed at 2 on 1

above a 5 on 1 slope to keep the cutting as narrow as possible, thus reducing the visual impact of the cutting. The road was opened in 1992. At tender stage it was anticipated that there would be areas of weak rock face, affected by faulting, which would require structural support. It was intended to provide a mesh-reinforced gunite face to these areas, following completion of the rock cutting. However, during the works it became clear that the areas affected were significantly greater than anticipated. The design of the road had attempted to limit the impact on the environment and consequently it was felt that significant quantities of mesh reinforced gunite would detract from the final appearance of the cutting faces.

Ove Arup and Partners were requested by the Client to provide a solution that would provide the necessary structural support and weather protection to the areas affected but which would also follow the structure and relief of the rock. In general the rock face was a fairly smooth surface formed by presplit blasting. In the weaker areas the blasting had formed a rough surface, typically recessed from the rest of the rock face, and this texture appealed to the clients landscape architects.

Following detailed research, SFC was recommended to the client and trials were carried out. It was considered that SFC would offer the following advantages over gunite:-

- the general relief of the rock face would not be lost;
- equivalent strength and durability would be achieved using less thickness and without relatively rigid mesh;
- there would be quicker construction and less material wastage.

The purpose of the SFC at Rhuallt was twofold. The main purpose was to provide a weather resistant coating to prevent erosion of the rock face. Secondly, it was required to provide structural support and this was aided by 3m rock bolts drilled into the weak rock.

1.2   Case study 2 : Stores Cavern, Dudley Zoo

Stores Cavern is an old limestone mine beneath Dudley Zoo with three access tunnels into the cavern which is up to ten metres high. It is proposed to open the cavern to the public for exhibitions. The roof of the cavern follows the bedding of the limestone, dipping at approximately 40°. However, the roof was undergoing progressive spalling failure due to thinly bedded limestone slabs separating from the beds above on the hanging wall of the mine, Figure 1. In addition, at the top of the hanging wall there is a bedding shear plane, dipping into the cavern and this was causing local instabilities. Traditional mesh reinforcement sprayed concrete was considered to be too dangerous to install as the attaching of the mesh could initiate failure of the already unstable roof. Following the successful use of SFC at Rhuallt it was recommended to the client that again SFC would be an ideal form of treatment. Unlike Rhuallt the main purpose of the SFC was to provide structural support. Consequently, it was recommended that the entrance tunnels, roof and high wall of the cavern be sprayed with SFC and supported by a grid of rock bolts at close centres.

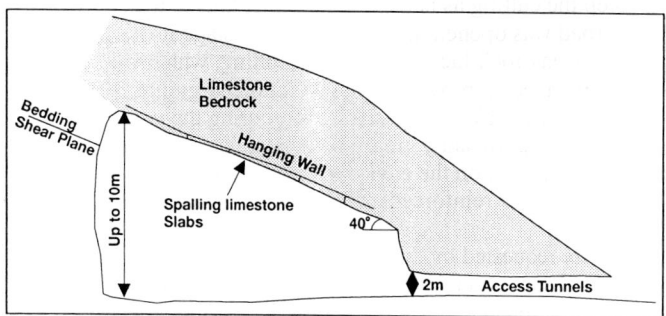

Figure 1 : Cross Section Through Stores Cavern

2 Design

SFC has been extensively used in America and Scandinavia since the late 1960s. Research in this country, mainly at Loughborough University, has been the basis for its gradual introduction into the UK, notably in the fields of slope stabilisation, underground support and repair work.

Research indicated that the inclusion of fibres in concrete improves its post crack ductility, toughness, impact resistance, fatigue resistance and most importantly, the flexural strength. It was found to have a superior load carrying capacity to conventional mesh reinforced sprayed concrete, especially at small deformations after first cracking and equivalent performance at large deformations, Figure 2.

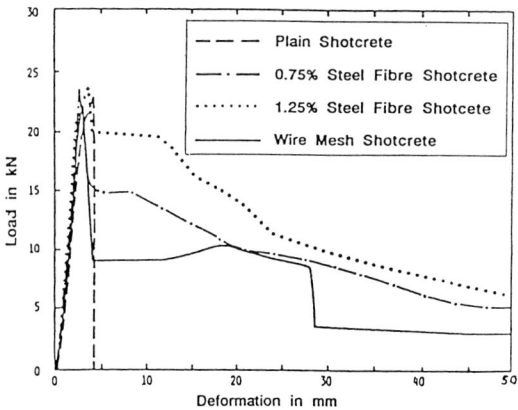

Figure 2 : Load vs Deformation for trial sprayed concrete panels (copied from [2]).

- Melt extract stainless steel fibres, 25mm long by 0.5mm (aspect ratio 50). It was considered that these fibres would provide good workability and relatively low rebound properties.
- In-situ fibre content not less than 2% by weight of the dry constituents. 3% was allowed for in the dry mix to allow for fibre lost due to 'rebound' during spraying.
- This mix should provide a flexural strength of $4MN/m^2$ and a toughness index of 10.
- In-situ thickness of 60mm, at the Stores Cavern a flash coating of 15mm was added to cover the fibres.
- The mix was a 4 to 1 mix of Class M aggregate to BS 882 and ordinary Portland Cement.
- Dye to be added to the mix to match natural rock colour.

Construction and quality assurance

The 'dry mix' method was used, in which the mix is blown by compressed air from the gun to the nozzle, where water is added. During trials various amounts of dye were added to the mix to match the colour of the rock. The trials demonstrated that SFC could be applied more quickly and in a more aesthetically pleasing form than conventional mesh-reinforced sprayed concrete.

To ensure that the design was being complied with, the following procedures were carried out:

- Samples were taken twice daily from the batching plant and the fibre content obtained by abstracting the fibres from the mix.
- Samples were taken twice daily from freshly sprayed SFC and the fibres extracted to obtain the weight in the mix, Figure 3.
- Samples were sprayed onto wooden boards each day and cored for compressive strength testing at 7 and 28 days. The specified strength of $30N/mm^2$ was usually achieved after 7 days.
- At Rhuallt there was insufficient time to carry out flexural strength testing as described in [3]. Consequently, cores were taken from the trial panels for point load testing from which the flexural strength of concrete can be determined, by the methods described in [3].

The results indicated an average flexural strength of $4MN/mm^2$, which was the anticipated design flexural strength.

At the Stores Cavern, using the same mix design, flexural strength tests were carried out on three samples cut from a single test panel. The tests were carried out in accordance with [4], the maximum flexural strength ranged between 5.2 to $7.5MN/m^2$ and the recorded mid-span deflection at first crack/failure ranged from 0.83mm to 1.15mm, see Table 1. These results indicate significantly higher flexural strengths than those obtained by point load tests described in [3]. However, the panel tested was thicker than the design thickness.

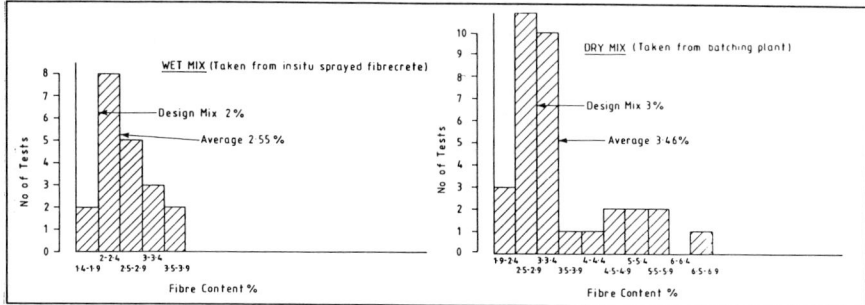

Figure 3 : Results of Fibre Content Tests

The wet mix samples were generally taken from the panels sprayed with a corresponding dry mix sample. So although there is a wide variability in values from the dry mix these values were never mirrored by the wet mix results suggesting further mixing in the spraying process.

Table 1 : Flexural strength test results of fibre reinforced concrete

Samples Ref	Test Span (mm)	Load at First Crack (kN)	Maximum Load (kN)	Sample Width (mm)	Sample Depth (mm)	First Crack Strength (N/mm²)	Deflection at First Crack (mm)	Maximum Flexural Strength (N/mm²)
C6/1	400	18.8	18.8	158	117	5.2	0.96	5.2
C6/2	400	29.2	29.2	155	123	7.5	1.15	7.5
C6/3	400	19.3	19.3	164	106	6.3	0.83	6.3

At both Rhuallt and Stores Caverns, Cores were also taken through the SFC and into the rock face to demonstrate the following:

- rock/SFC bond;
- density of SFC;
- SFC/SFC bond where more than one layer had been sprayed;
- fibre distribution

These were assessed by visual inspection, no bond strength tests were carried out.

The results of the coring indicated that the rock/SFC and SFC/SFC bonds were generally good, even in badly faulted ground. In addition, the density and distribution of fibres was good, with the fibres orientated parallel to the sprayed face. Prior to spraying SFC, trials were undertaken to determine whether compressed air or high pressure water jet was the most suitable method for cleaning the rock faces. A high pressure water jet was used in highly faulted areas, so that the rock face would be slightly recessed prior to spraying. In weathered areas, compressed air was used to remove any loose debris. Cleaning was conducted from the top down, to ensure that no loose material was left on the face.

Drainage
In the faulted areas at Rhuallt Hill, where water flows were high, large diameter drainage holes (100mm) were drilled to lengths of between 6 and 12m. These were generally drilled upward at an angle of 10° and slotted plastic pipes were installed in the holes.

During spraying of the SFC, at both Rhuallt and the Stores Cavern, small diameter drainage holes were installed at intervals at a rate of 1 per 2m². Following spraying, if water was still observed to be seeping through the SFC, additional holes were drilled to relieve the water pressure.

Structural support
At Rhuallt the sprayed areas were generally limited to areas of weak/weathered rock and were typically up to 12m high, with lengths up to 20m, but generally 5 to 10m. Consequently, it was considered that it would be prudent to install rock bolts into the face to limit the span of the SFC; 3m long rock bolts were installed typically on a 3m grid.

At the Stores Cavern, the SFC was generally sprayed onto the hanging wall of an old mine undergoing spalling failures. In this case, rock bolts were primarily needed to stitch the beds of limestone together, to prevent further deterioration. The second function of the bolts was therefore to provide support to the SFC.

Monitoring
At the time of the works the SFC at Rhuallt was considered to be the largest single external application in the UK. Consequently, it was recommend that the faces of Rhuallt be monitored. This was to entail:-

- records of cracking.
- identifying debonded areas using hammers.
- recording drainage and areas of seepage and providing new drainage if required.

Monitoring to date has only revealed minor cracking in the SFC. This occurred shortly after application and may therefore have been due to shrinkage despite the concrete being sprayed with a curing agent.

Specification
Following the Rhuallt and Stores Cavern jobs where the obvious advantages of SFC over traditional mesh reinforced sprayed concrete were observed. It was decided to introduce a specification into the Ove Arup system to encourage further use of this material.

The specification covers such items as:-

- Materials to be used, aggregates, fibres, cement, etc.
- Plant
- Qualification of Operators
- Surface preparation and spraying procedure
- Design and Quality Control

Further guidance concerning specification of such materials is given in [5].

Conclusions

The purpose of mesh in sprayed concrete is primarily to limit cracking. SFC, has the following advantages over conventional mesh reinforced sprayed concrete:-

- better stress-strain properties, post cracking;
- savings in costs of fixing mesh, leading to faster construction and reduced time on site;
- reduced design thickness;
- reduced volume of material required to cover uneven surfaces;
- superior aesthetic qualities to mesh reinforced sprayed concrete, which produces a featureless surface in stark contrast to the remainder of the rock face.

Both case studies have illustrated the advantages of SFC over conventional mesh reinforced sprayed concrete in civil engineering schemes. At Rhuallt it provided an aesthetic solutions in an environmental sensitive scheme. In addition, the vast quantities needed, up to 1,500m^2, were installed in a shorter timescale than conventional mesh reinforced concrete, giving less disturbance to other site activities.

At Stores Cavern, SFC provided a quicker and safer solution than conventional meshing and guniting, with no loss of structural performances.

From the experiences of the author, SFC is a material with many advantages over traditional sprayed concrete and with many potential applications. However, only further recording of the use of such materials and the production of relevant design guides and specifications will encourage its use.

REFERENCES

1. Concrete Library of JSCE No.3, June 1984, *Recommendations for Design and Construction of Steel Fibre Reinforced Concrete.*
2. Morgan, D.R. (1988) *Recent Development in Shotcrete Technology - A Materials Engineering Perspective*, Presented at World of Concrete, 1988, Las Vegas USA.
3. Robins, P.J. and Austin, S.A. (1985) *Core point load test for steel fibre reinforced concrete*, Magazine of Concrete Research : Vol 37, No. 133, December 1985.
4. ASTM *Standard Method of test for flexural toughness and first crack strength of fibre reinforced concrete*, ASTM C1018, April 1985.
5. Austin, S. and Robins, P. (Eds) (1995) *Sprayed Concrete, Properties, Design and Applications*, Published by Whittles Publishing.

PART FIVE
REPAIR

16 THE REPAIR OF GOREY JETTY IN THE STATE OF JERSEY

K. Armstrong
Public Services Dept, States of Jersey, St Helier, Jersey, CI
P. Quarton
Connaught Group Ltd, Gloucester, UK

Abstract
The repair of marine structures contaminated by chloride ingress can be a complex task involving the understanding of survey information, formulation of a repair regime and selection of appropriate materials, for an engineered solution within budget guidelines. This paper describes the processes followed to repair a reinforced concrete jetty subjected to high tidal action and the use of electrochemical desalination to extract chlorides from the concrete prior to application of a sprayed cementitious overlay.
Keywords: Concrete, chloride ingress, corrosion, marine, tidal, reinforcement, sprayed concrete, desalination, repairs.

1 Introduction

Gorey Jetty, is located at the southern end of Gorey Pier (Fig. 1), on the east coast of Jersey, Channel Islands. Situated within the Bay of St. Malo on the north coast of France, Jersey is subjected to a tidal range of 12.2m (40ft) reported to be the fourth highest in the world.
The structure, constructed in 1978/79 in reinforced concrete, is founded on fill (Fig. 2). The concrete frame is made up of precast columns and beams with insitu slab and beam infill between frame units. The Jetty is built adjacent to an older masonry harbour wall.

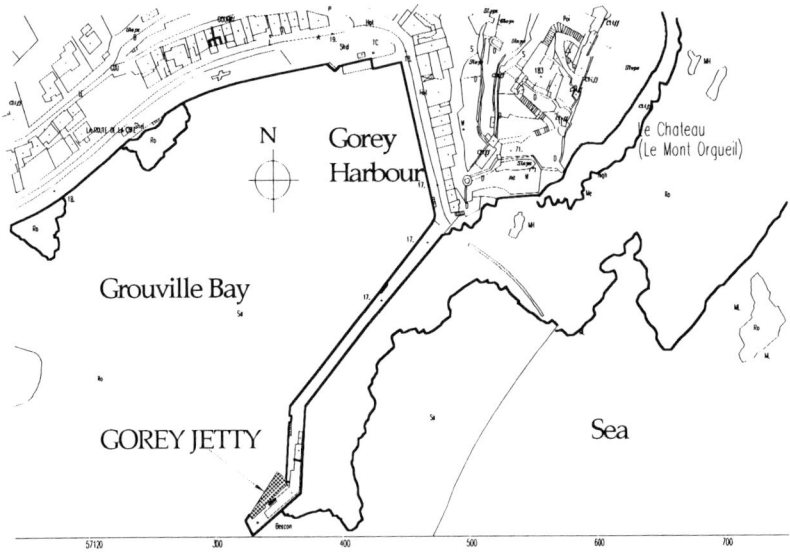

Fig. 1. Location plan

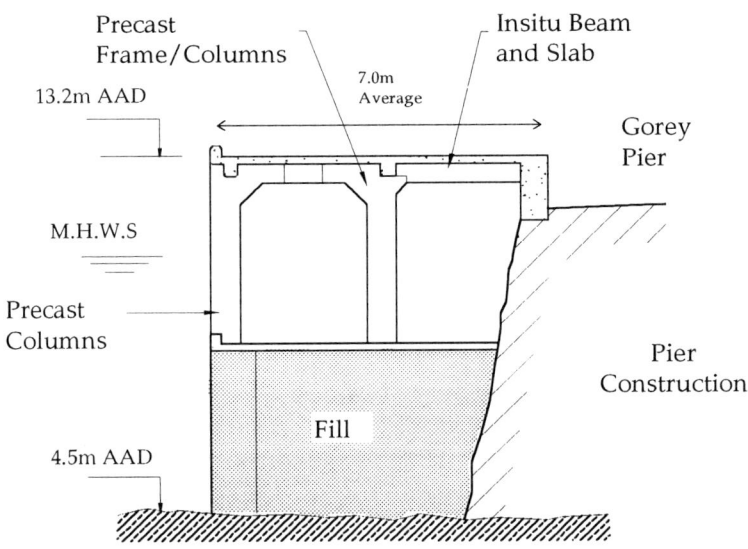

Fig. 2. Section through Gorey Jetty

2 Survey

A visual concrete condition survey highlighted spalling and numerous cracks with rust staining and surface pop outs. A half cell, covermeter and concrete dust sample survey were commissioned with petrographic analysis of the concrete to assess the extent of degradation. This confirmed that all the concrete examined exhibited reinforcement corrosion precipitated predominantly by excessive externally derived chloride.
The original drawings indicated the concrete as being Grade 30 OPC.

2.1 Covermeter survey
Minimum depths of cover ranged from 20 mm to 33 mm for the precast concrete columns and from 26 mm to 57 mm for the insitu cast beams. Slab cover depths varied between 31 mm to 55 mm for the insitu areas and over 50 mm for the inclined staircase soffit.

2.2 Carbonation tests
By phenolphthalein testing on site, carbonation readings of 5 mm to 15 mm for the insitu beams and slab and less than 5mm for the precast elements were found. These results were corroborated by microscopy on cores taken in sample areas.

2.3 Chloride tests
Dust samples taken at three depths, 5 mm to 30 mm, 30 mm to 60 mm and 60 mm to 90 mm. Values of Chloride ion content expressed as % by weight of cement assuming a cement content of 14% are shown in table 1. The chloride ion profiles were found to significantly decrease inwards, although levels from the 60 mm to 90 mm were still in the medium to high risk category as defined in the BRE Digest 264 [1].

2.4 Half cell potential testing
Half cell potential testing was carried out in accordance with ASTM C876 [2], using copper-copper sulphate half cell. Results are shown on table 1. Results greater than -350 mV are considered to have a greater than 90% probability that corrosion is occurring but corrosion activity for results in the range -200 mV to -350 mV are considered as uncertain.

2.5 Petrographic examination
The petrographic examination of the concrete cores revealed that both the precast and the insitu concrete are of similar composition. Coarse and fine aggregates are of a hornblende gabbro and diorite with some granitic material. The fine aggregates also contain some natural quartz sand. Some potentially deleterious materials were also in evidence, namely strained quartz and micro-crystalline quartz. Evidence of alkali aggregate reaction in the samples was classified as only "trivial". Minor aggregate shrinkage was also observed.

The cement matrix of the concrete sample showed a high microporosity, possibly the result of an estimated high water/cement ratio of 0.62. This relatively open matrix would allow easy moisture and chloride ingress. Carbonation would normally advance rapidly but the intermittent wetting, in the tidal environment, blocks the passage of carbon dioxide through the concrete.

The approximate concrete strength ranged between 34 N/mm^2 to 44 N/mm^2.

Table 1. Summary of test results

Test Area	Element	Min. Depth Cover (mm)	Max. Depth Carbonation (mm)	Chloride Content %	Half Cell Potential (mV)
1	Insitu Beam	31	10	0.43 - 1.09	-110 to -344
	Precast Beam	Not tested	5	0.10 - 0.99	Not tested
2	Precast Col.	26	<1	0.22 - 1.57	-190 to -437
3	Insitu Slab	34	5 - 10	0.51 - 1.52	-93 to -348
4	Staircase	50+	<5	0.51 - 1.66	-214 to -378
5	Precast Col.	20	<5	0.19 - 1.52	-183 to -437
6	Insitu Beam	27	5 - 10	0.58 - 2.60	-247 to -462
7	Insitu Beam	57	<5	0.31 - 1.50	-254 to -346
8	Precast Col.	33	<1 -<5	0.43 - 1.47	-246 to -516
9	Insitu Beam	33	10 - 15	0.31 - 2.05	-232 to -344
10	Insitu Slab	31	10	0.22 - 1.37	-107 to -268
11	Insitu Beam	26	15	0.48 - 1.88	-230 to -348
	Precast Col.	31	<1	0.12 - 0.41	-180 to -216
12	Insitu Slab	49	5 - 10	0.80 - 2.07	-192 to -318
13	Precast Col.	29	5	0.10 - 1.91	-68 to -181
14	Precast Col.	30	1	0.05 - 1.54	-216 to -519
15	Insitu Slab	55	<5	0.27 - 0.53	-48 to -272
Core Samples					
1	Precast Col.		1.5	Not tested	
2	Precast Col.		1.0	Not tested	
3	Precast Col.		5.0	Not tested	
4	Precast Col.		Not tested	0.07 - 1.29	
5	Precast Col.		Not tested	0.07 - 1.64	
6	Precast Col.		Not tested	0.07 - 1.64	
7	Insitu Beam		14	Not tested	
8	Insitu Slab		16 top & 8 soffit	Not tested	

3 Appraisal of results

From the survey information the main cause of deterioration to the concrete is by chloride ion ingress and subsequent corrosion of the reinforcement leading to spalling of the cover concrete. Chlorides attack the passive layer surrounding the reinforcing bars in alkaline concrete and this action results in highly active corrosion to localised areas. If left unchecked the corrosion will lead to loss of section of reinforcement and ultimately the possible collapse of the structural element.

With the chloride ion levels above 0.4% at depth of reinforcement and values increasing to the surface and decreasing with depth, the source of the chlorides was considered as being externally derived.

Half cell results indicated that a majority of the structure elements were in the high risk category with other areas at a relatively passive stage, prior to corrosion starting. Given the chloride levels within the concrete, the time for activation of corrosion could be considered as imminent.

The depths of cover and carbonation depths were at acceptable levels.

The identification of "trivial" amounts of alkali aggregate reaction was ignored in the light of the severity of the other forms of degradation encountered.

4 Repair Options

Any repair option must address the cause of the problem by preventing further degradation, or at least slow the processes down to an acceptable, manageable level.

The above statement could be refined by the inclusion of a monetary limit, ... and **to be cost effective, subject to budget requirements.**

From the survey results the causes and effects of degradation are reported as:-

Cause
- Chloride ingress
- Porous nature of concrete
- Low cover

Effect
- Spalling
- Corroding reinforcement
- Surface staining
- Cracks

So the ideal repair method for the Jetty should remove the chlorides from the concrete, restore cover and inhibit remigration, or prevent the reinforcement from becoming actively corrosive. The methods investigated are given in table 2.

Table 2. Repair options.

Repair Option	Advantages	Disadvantages
Patch repairs.	Relatively inexpensive. Short contract programme.	Only spalls cut out. Chlorides still allowed to cause corrosion. Patchy appearance. Short lifespan.
Patch repairs with sprayed concrete overlay.	Additional cover to rebar. Barrier to chlorides. Compatible material to base concrete. Uniform colour. Durable overlay.	Only spalls cut out. Additional load to structure. Preparation required to all areas. Medium cost.
Patch repairs with non-cementitious overlay.	Barrier to chlorides. Uniform colour. Crack bridging properties.	Product selection ? Longevity ? Water vapour transmission rate non-compatibility. Surface porefilling required. Preparation crucial.
Desalination + Sprayed concrete overlay.	Reduces chlorides. Barrier to further chlorides. Enhanced lifespan.	Expensive. Unproven in tidal zone. Additional load to structure. Preparation crucial. Messy.
Cathodic Protection + repairs.	Enhanced lifespan. Inhibits corrosion.	Expensive. Maintenance costs. Anode replacement.

In consideration of the repair options, the use of Cathodic Protection was not investigated further owing to a perceived initial high cost and ongoing monitoring. Just patch repairs were discounted as an option, they provided no barrier to chlorides ingress. The question of desalination of the whole structure and its effectiveness for concrete that is continually wetted by the tide was subject to trials being carried out on another structure in Jersey [3]. The process had proved workable and suitable for deck slabs above high water and would be considered for the Jetty soffit, if the costs were within budget.

Non-cementitious overlays were not considered further due to the lack of a proven product on the market that can withstand the aggressive environment and be applied to wet concrete.

Unfortunately it became apparent that the budget for this project would not be sufficient to effect a full repair by removing the cause of the corrosion. The second best option would therefore be to remove the degraded areas and reinstate the spalled concrete and slow down the chloride ingress.

Patch repairs with a sprayed concrete overlay offered a barrier to further chloride ingress but did not remove the potential for chloride attack from the chlorides locked in the concrete. Indeed removing areas of spalled concrete can move the corrosion cathode sites to new areas previously unaffected. This potential for continuing corrosion activity needed to be borne in mind before a repair regime was finalised.

Uncertainty as to the cost of patch repairs with a sprayed concrete overlay on such a structure, led to a tender being sought by Public Services Department for this basic system. The tender return proved the economic viability of sprayed concrete and left sufficient funds remaining in the budget to enhance the repair regime. The use of desalination to the soffit of the top deck proved to be feasible after analysis of the survey data relating to chloride contents, cover to rebar and reinforcement location and quantity.

The question of removing chlorides or inhibiting corrosion to the precast columns was discussed and will be addressed at a follow up contract, subject to monies being available in 3 to 4 years time. Although this decision is not an ideal situation from an engineering point of view, it does buy time, time to investigate and consider alternative forms of repair.

5 Material selection

The specification of repair materials is never a straight forward one when dealing with a tidal structure. The aggressive environment as shown in Fig. 3. calls for specialist materials to combat the conditions. The sprayed concrete needed to withstand the effects of a rising tide (washout) some 3 to 4 hours after application. Sikacem 133 Gunite (a cement based, polymer-modified one component repair mortar containing silica fume and high range water reducing agents) had successfully been used elsewhere in Jersey and was chosen for this project. A 20 mm thick overlay would be applied to all concrete surfaces. The structure was considered capable of taking the additional load.

Waterproofing to the top of the concrete deck is essential to prevent water ingress into the slab. The concrete slab is subjected to foot traffic only. Several products were available ranging from cementitious screeds to simple coatings. Any cracks in the surface are being treated separately by chasing and filling with a sealant or filled with concrete repair materials depending on whether they were active or non active cracks. As the coating was not required to have any crack bridging properties, a hydrophobic impregnant with an low viscosity epoxy impregnation material was chosen, namely Icosit Aquastop overcoated with Sikafloor 619.

Concrete repair materials to repair areas above the tidal zone were chosen from the Sika range, Sikatop 121 levelling coat and an anti-carbonation coating Sikagard 680S.

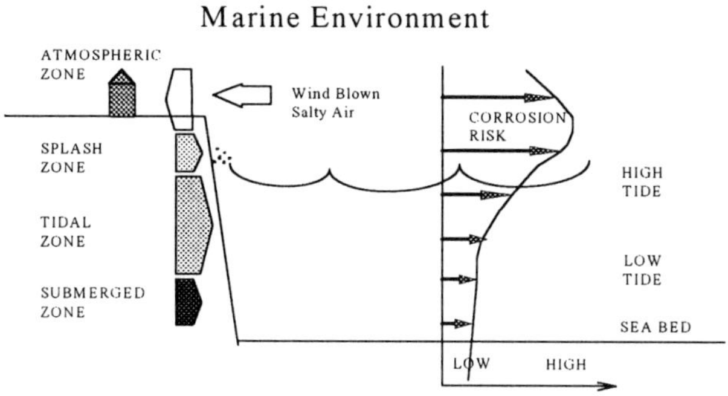

Marine Environment

Fig. 3. Corrosion risk to reinforcement in marine exposure zones [4]

6 Site operations

Connaught Structural Repairs Limited won the contract and work started in February 1996. The use of hydro-demolition of concrete was a first for Jersey and proved to be an economic and environmentally efficient way of removing defective concrete. The water is pressurised to 15000 psi at the nozzle end and removes concrete by cutting or a bursting action. This method proved invaluable when the decision to remove five reinforced concrete beams was taken due to the severity of cracking encountered.
These beams also lacked the prescribed amount of reinforcement as detailed on the original drawings and even more worrying was the misplacing of bars at joints. Upon examination the extent and nature of the cracking in these elements could have indicated that an early age movement had occurred to the structure by a unknown external force. The beams were recast using SBD five star flowable repair concrete.

The desalination of the top deck soffit was divided up into two sections, one section desalinated at a time, each section consisting of bays totalling similar surface areas. Shutters, bolted to the concrete soffit in each bay, contained a solution of calcium hydroxide and water together with titanium mesh and connected to a power supply unit (RD). A maximum of 30 volts at 10 amps can be passed through the cables. The actual supply is determined by the resistivity of the concrete and the surface area of the reinforcing bars in contact. A computer monitors the supply and resistance offered by the concrete to ensure control of the process.

Dust samples were taken from the concrete before, during and after the process, including analysis of the chloride ion content of the electrolyte. The samples were analysed by a independent testing house for chloride ion content. The results from these tests determine the length of time that the desalination runs, on average the process took four weeks. A target level of chloride ion content was set at 0.5% - 0.6% by weight of cement.

All concrete surfaces were prepared by grit blasting prior to application of 20 mm of Sikacem 133 gunite. This material was applied using dry spray equipment. A natural sprayed concrete finish was specified to all concrete surfaces apart from the columns where a trowelled finish was used to lessen the extent of marine growth and water borne detritus becoming attached to the surface.

Repairs to spalled concrete were also sprayed. The rebar being grit blasted immediately before filling.

The contract is due for completion in mid June 1996, the programme being wholly dependant on the time for the desalination process to be completed. The actual time for the basic contract, excluding the desalination, would be have been approximately nine weeks.

7 Conclusions

- Hydro-demolition offers an economic and efficient method of cutting concrete.
- Sprayed concrete can be effectively applied, as a barrier to further chloride ingress, to concrete in the marine environment. It also provides additional cover to reinforcement.
- Desalination with a sprayed concrete overlay will restore the structure to a passive condition giving an extended life over the non-repaired state. Careful analysis of the suitability of desalination for a structure is required.
- Gorey Jetty has been partly restored to a passive condition. A barrier to chloride ingress has been applied to all concrete surfaces. The column areas can now be assessed for additional forms of treatment when funds are available.

8 References

1. Building Research Establishment.(1982) *The durability of steel in concrete: Part 2. Diagnosis and assessment of corrosion-cracked concrete.* BRE Garston UK. BRE Digest 264.
2. ASTM Committee on Standards. (1989) *Test method for half-cell potential for uncoated reinforcing steel in concrete.* ASTM Philadelphia. ASTM C 876-89, Annual book for ASTM Standards.
3. Armstrong, K.,Grantham, M.G., McFarland, B. (1996) The trail repair of Victoria Pier, St Helier, Jersey using electrochemical desalination. *Procs. of Corrosion of reinforcement in concrete construction.* Cambridge.
4. Comité Euro-International du Béton. (1992) *Durable concrete structures.* CEB, Lausanne. Design Guide.

17 WET PROCESS SPRAYED CONCRETE TECHNOLOGY FOR REPAIR

S.A. Austin, P.J. Robins, J. Seymour and N.J. Turner
Dept of Civil & Building Engineering, Loughborough University of Technology, Loughborough, UK

Abstract
This paper describes a research programme aimed at advancing the understanding and technology of the wet process to enable it to be specified and used with confidence for repair applications in the United Kingdom. In particular it describes the results of an industrial survey on repair contract characterisitics and initial trials on the workability (the two-point shear vane type) and pumpability of several mixes. The results demonstrate the low dependance on operator skill and the crucial effect of the aggregate grading with rotor stator type equipment.
Keywords: repair, rheology, spraying process, wet mix, wet process

1 Introduction

Currently, sprayed concrete repair projects in the UK are almost exclusively carried out by the dry process. The wet process has become dominant for large scale tunnelling applications involving robot-controlled spraying (e.g. in Scandinavia and more recently in the UK with the New Austrian Tunnelling Method or NATM), but is not a common solution for repair work.

The dry process is capable of producing high quality concrete but has several drawbacks including quality and consistency, high rebound losses and a dusty and dirty environment. The wet-mix process has the potential to produce more consistent concrete with lower rebound, at lower cost and in a healthier working environment, but much of the technology developed to date is inappropriate for repair work, because it is based on rock support, inappropriate mixes and high volume production (not controlled overlays).

The main emphasis of our work is on mortars and small aggregate concretes (<10 mm) applied in thin layers (<100 mm) at controlled low/medium output rates

($<5m^3/hr$), in some cases with mesh or fibre reinforcement. The wet process could replace a significant amount of current dry-mix work and also significantly extend the use of sprayed concrete in repair work, the latter involving expenditure of around £500 million per annum in the UK. Whilst sprayed concrete material's costs would only account for a fraction of this total, the potential savings are much greater if more durable repairs can be effected which increase the remaining life expectancy of the structure.

The main objectives are:
1. to gain a fundamental understanding of the influence of the pumping/spraying process, mix constituents and proportions on the fresh and hardened properties of wet-mix sprayed concrete;
2. to improve the wet-mix spraying process, in particular operator environment, maximum conveying distances and stop-start flexibility;
3. to specify, measure and optimise in-situ properties, particularly strength, bond and durability;
4. to disseminate information in appropriate form to practising engineers to promote and accelerate the use of wet-mix sprayed concrete for repair in the UK.

This paper reports on work related to the first objective.

2 Current position in wet and dry process sprayed concrete

In the United Kingdom sprayed concrete repair projects are almost exclusively carried out using the dry process, with a recent example being the repair to the Queensway Tunnel in Liverpool [1]. The underside of the road deck required repair following corrosion as a result of chloride ingress and carbonation. Another example is the repair of the Runcorn Bridge over the River Mersey [2], where dry sprayed concrete was required to protect the titanium mesh which formed part of the cathodic protection system.

The main concern regarding shotcrete for repair at present is the durability and compatibility of the repair material with the substrate, on which research is currently being carried out at many institutions in North America and Europe. Tests methods for predicting the durability of repairs are also being developed and existing ones are being improved.

The dry process is capable of producing high quality concrete but has three significant drawbacks:
1. quality and consistency are a function of water content, operating pressures and spraying techniques, all of which are highly dependent on the skill and care of the operatives;
2. high material losses (20 to 40% on vertical faces, 30 to 50% overhead), consisting largely of rebound aggregates, are not only wasteful but present a substantial removal problem and can get trapped behind reinforcement (introducing weaknesses); and
3. spraying produces a dusty and dirty environment, which can be harmful to operatives and can cause problems on sensitive sites or in restricted spaces.

The wet mix has the potential to produce more consistent and controllable concrete with lower wastage (5 to 10%), at a lower cost, and in a much healthier working environment. However, much of the technology developed to date is inappropriate for sprayed concrete repair work in the UK, because it is based on rock support, inappropriate mixes and high volume production. The disadvantages include shorter transporting distances of the pumped concrete, difficulty in the handling of a heavy, concrete-filled hose and nozzle, and a lack of flexibility. Where there are numerous breaks in placement, it is not suitable to process small quantities of sprayed concrete, as prolonged stoppages in spraying necessitate the removal of hardening concrete, and a through cleaning of the spraying unit.

In some countries there has been a big swing towards the wet process, partly because of better control over mix proportions (particularly the water/cementitious materials ratio). These include Norway and Sweden, where the majority of work is wet process, and the USA where the two techniques have a roughly equal share and are both used for repair [3].

Although the proportion of wet sprayed concrete is increasing in the UK, other countries (in particular Germany) are still predominately orientated towards the dry process. These differences partly reflect the functional emphasis of the two processes (e.g. wet mix for high output applications such as tunnelling, and dry mix for low to medium output applications such as repair, or situations requiring greater transport distances and flexibility like mining).

3 Concrete Repair Scenarios

A survey of local authorities, consultants, contractors and material suppliers was carried out in order to identify a set of generic repair scenarios and their performance requirements, to cover the broad range of repair situations encountered in the UK [4].

Eleven organisations participated in the survey, each providing information either through interviews or by completion of a standard questionnaire form. The scenarios are classified in terms of characteristics common to various repair applications, including; purpose; orientation; geometry; reinforcement; substrate; surface finish; construction method and environment.

Table 1 contains the repair scenarios generally encountered in the UK at present. From this it can be seen that four main categories of structure require repair; these are R.C. bridges, R.C. buildings, R.C. tunnels, and masonry structures. The majourity of repairs at present are to motorway bridges, with repairs to high rise buildings being the next most common.

The main causes of deterioration of concrete leading to repair are the corrosion of steel reinforcement due to the ingress of chlorides and carbonation, alkali aggregate reaction and the effects of heat following fires. Typical repairs are normally below 2 m^2 in size with a depth of 50 to 100 mm. At present most bridge repairs are carried out using flowable concretes, with repairs to high rise buildings mainly carried out by applying hand packing concrete. A smaller proportion of each are carried out with dry sprayed concrete, particularly when the size of repair is above $1m^2$.

Table 1. Repair Scenarios for investigation using wet process sprayed concrete technology.

TYPE OF REPAIR DESCRIPT.	PURPOSE	GEOMETRY SIZE	GEOMETRY DEPTH	GEOMETRY TOL.	PREPARATION	SUBSTRATE TYPE	SUBSTRATE SURFACE CHAR.	ENVIRON.	ORIEN.	REINF.	SURFACE FINISH	ACCESS
bridge soffit	1 cover 2 structural	<2m³	50-100mm	+/-10mm	hydrodemolition + grit blasting	concrete	AAR Carbonation Chlorides	atmos.	overhead	mesh fibre	troweled finish, no colour match	limited to night restricted space available.
bridge abutment (marine structures)	1 cover 2 structural	<2m³	50-100mm	+/-10mm	hydrodemolition + grit blasting	concrete	AAR Carbonation Chlorides	atmos.	vertical	mesh fibre	troweled finish, no colour match	limited to night restricted space available.
building (water retaining structures + r.c. chimneys)	1 cover 2 structural	<2m³	50-100mm	+/-10mm	mechanical hydrodemolition + grit blasting	concrete	carbonation (chlorides in car parks)	atmos.	60:40 vertical: overhead	mesh	troweled, colour match (where no surface coating provided)	external repairs use scaffold, platforms etc.
Fire-damaged structure	structural	<2m³	50-100mm	+/- 3mm visible +/- 10mm covered	hydrodemolition + grit blasting	concrete	fire-damage	atmos. (substrate absorbs H2O at high rate)	50:50 overhead: vertical	mesh (replace damaged steel)	troweled where visible, otherwise as shot, no colour match	ok
tunnels	structural	<1m³	100mm	+/-10mm	hydrodemolition + grit blasting	concrete	carbonation chlorides	cool (ventilation fans) can be damp	overhead	mesh, corroded steel replaced	as shot, no colour match	restricted access to road and rail tunnels, pumping long distances
sewer (masonry tunnels + arch bridges)	strengthening	1m³+	25-50mm (less than 100mm)	+/-10mm	grit blasting	masonry	deteriorated masonry	warm & damp	curved surface	mesh used (stainless steel)	as shot, no colour match	restricted access through man holes, pumping long distances

NA Not Applicable

It was found that most repairs are carried out with proprietary prebagged materials/products from manufacturers. These are used because they are perceived to be of a higher quality than site batching (in terms of the quality assurance of the ingredients) or ready mixed concrete. Interestingly, contractors and local authorities carry out few, if any, quality control tests on these prepacked materials, considering it unnecessary and too costly

From the survey it was found that the orientation was divided equally between overhead and vertical applications. However, access to the repair sites can often be difficult, with the space and time allowed to carry out the repair being limited. The information obtained from this survey is being used to investigate the potential for using wet process sprayed concrete technology for the repair scenarios identified.

3 Mix design

A range of base mix formulations were needed to act as a starting point for the research project. For an effective repair process it is likely that each scenario, or groups of scenarios, will require different mix designs as a result of their varying characteristics. The performance criteria of the scenarios will form targets and acceptance criteria, against which the mixes and equipment set-ups under development will be evaluated later in the project.

An extensive review of published literature and current practices on materials requirements and mix design parameters concerning concrete for pumping and spraying was conducted. A simple mix design was chosen from these examples, allowing the basic materials to be tested, and then developed during future trials to suit the different applications. A ready mix design for pumped concrete was adapted to produce a mortar mix with an aggregate/cement ratio of 3:1, thus giving batch quantities of $1613 kg/m^3$ and $537 kg/m^3$ respectively. The water content was determined in the laboratory to meet a target slump of 50-75 mm. This was to aid in the process of designing a series of base mixes for the initial laboratory pumping and spraying trials [5].

4 Pumping and spraying trials

A crucial aspect of successful wet-mix production is to achieve a balance between the characteristics of a pumpable mix, and those required to project it in to place with minimum losses, slump and segregation. In a preliminary pumping and spraying trial conducted at Loughborough University, a rheological study was made of the pumpability of various mixes using workability tests. Rheology is defined as the science of deformation and flow of matter, and covers relationships between stress, strain and time. In terms of fresh concrete, the field of rheology is related to the flow properties of concrete, or with its mobility before setting takes place.

Current standard tests for workability (slump, flow table, VeBe) tend to be inadequate, as they provide an estimate of the fundamental properties relating to rheology from a single result. The two-point shear vane workability test, developed by Tattersall [6] describes workability in terms of two constants. It assesses the flow properties of concrete by measuring the torque required to rotate a standard impeller immersed in concrete, at several different speeds. This is possible because concrete under test behaves according to the Bingham model:

$$T = g + hN$$

Where T is torque, N represents speed, the constants g and h together describe the workability of the concrete, g is a measure of the yield value and h is the plastic viscosity.

Initial trials were carried out using the two point test apparatus. The workability of 1:2.7:6.3 10mm concrete mix was varied by changing the water content. The results of this study are shown in Figure 1. From this it is clear that as workability increases the torque reduces for a specific impeller speed. It was also apparent that the apparatus does not perform well in the low workability range, i.e. high T values. At slower speeds in low workability mixes, a much greater force seems to be required to enable the impeller to rotate, thus producing seemingly anomalous results. However, if the results of these mixes below 0.5 rev/s are omitted (as they are in Figure 1), the results appear to be satisfactory. The low workability mixes shown, (w/c's of 0.5 and 0.55) correspond to slumps well below 50 mm, the recommended minimum for this apparatus. For this reason the apparatus should be satisfactory for trials on wet sprayed concrete which requires a slump of 50 to 100 mm.

Fig 1. Two-point workability results for the concrete mix

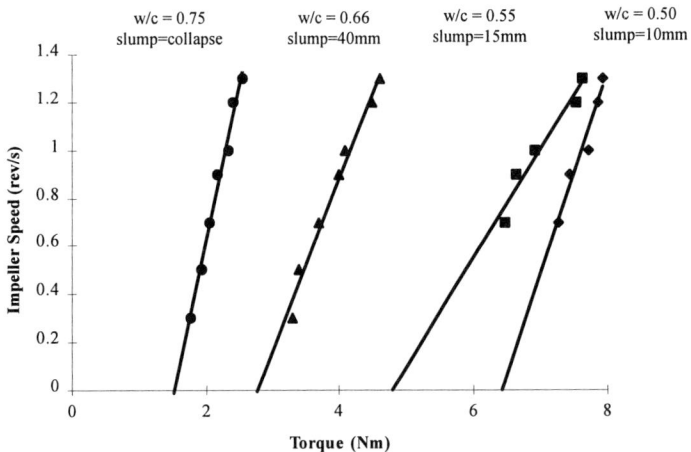

The first wet-sprayed concrete trial was conducted in January. Several sands were obtained from suppliers in the East Midlands, selected on the basis of grading curves, particle shape, and aggregate types. The aggregates used in the trial were a river sand having rounded and smooth particles, and a granite dust having angular and granular particles. The actual gradations of these aggregates are given in Table 2. The EFNARC specification [7] for the grading of aggregates for wet sprayed concrete was used as a guide.

Table 2. Gradations of aggregates used in spraying trial

Sieve size (mm)	River sand	Granite dust
10	100	100
5	99.7	99.3
2.36	90.3	81.6
1.18	75.3	60.2
0.6	57.2	42.8
0.3	21.4	26.8
0.15	0.02	13.4

The trial mixes investigated were:
- a) 6mm down granite sand/cement 3:1 mix;
- b) 4mm down river sand/cement 3:1 mix; and
- c) a proprietary material

The equipment used for the trial was a Concrete Repair Unit, designed to pump and spray a proprietary mortar mix, plus a 250 c.f.m. compressor. The unit was fitted with a 6 mm rotator stator, thus 6 mm maximum grade aggregate was chosen. Fresh concrete properties and characteristics such as pumpability and sprayability were examined. The strength of the materials were also evaluated using core samples taken from trial panels, and from cast 100 mm cubes. When further insight has been gained into the pumping characteristics of mortar and concrete, requirements relating more directly to various repair scenarios such as sprayability (build-up in vertical and overhead orientations) and hardened properties will be investigated.

The experimental procedure consisted of spraying trial panels, and taking samples of fresh mixed, pumped and sprayed material to conduct a two-point workability test at approximately the same time after mixing. Other measurements taken included output rate, slump, air and mortar temperatures, and compressive strength to be used later in the project (but not reported here).

The first material tested was a proprietary repair material, which, although designed for hand application, has been proven to pump and spray with the Concrete Repair Unit. Results of tests carried out on this and the other mixes are shown in Figure 2.

The workability results for proprietary material at each stage in the process might appear to be different to those expected, where progressive loss of workability occurs due to the effects of pumping and shooting compaction, as shown by Beaupré [8]. His results however, were obtained on air-entrained mixes, where significantly more compaction would be expected. Our results show that the pumped mix proved to be more workable than material taken directly from the mixer. The mix was sprayed three times, by an experienced operative and two researchers. The latter quickly learnt how to use the nozzle and produced very similar compressive strengths (29 and 31 Mpa compared with 32Mpa at 7 days).

Fig 2. Two-point workability test results from spraying trial

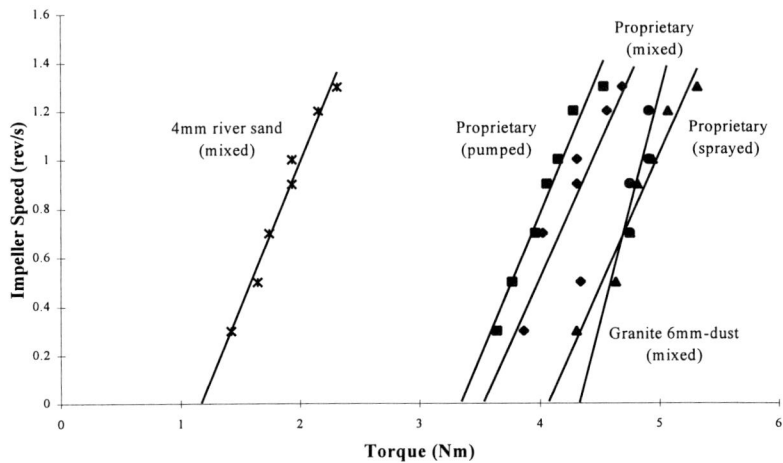

The 3:1 mix made with granite dust did not pump, and this appeared to be due to the largest particles jamming the rotor stator. A two-point workability test conducted on material taken from the mixer, shows the mix to be less workable than the proprietary blend, and perhaps approaching some limit in pumpability (Figure 2).

The 3:1 mix made with a sharp washed sand had been sieved to a 4 mm maximum particle size. The water content of the mix, previously estimated before sieving took place, had thus changed. The new water content was estimated by adding 100ml volumes at time; the mix appeared to be water sensitive, and a mixture of high workability resulted. However, this mix was also unpumpable due to a sand blockage, which is likely to have been caused by loss of water through gaps in the material grading, not jaming by the aggregate particles. Its low Torque also suggests the mix was too workable.

5. Conclusions and further work

The results obtained have demonstrated the critical dependence of rotor stator type wet process pumping on aggregate grading, which will have important implications for mix design and batching (site versus pre-blended). The trial has shown that a well graded material pumps and sprays readily, produces satisfactory compressive strengths and that the level of operator skill required is significantly less than for the dry process.

The results obtained from the two point workability tests were promising as a measure of pumpability and possibly defining workability ranges for the wet process.

A second trial is being planned for the near future, with different materials, and building on the invaluable knowledge and experience gained from the first trial. Further work is also being conducted to determine the limits of the two point workability testing apparatus in terms of low workability mixes.

Acknowledgements

We would like to acknowledge the financial support of the EPSRC and to thank FOSROC International (one of the projects industrial collaborators) for their help with the pumping and spraying trial, and the organisations who assisted in the survey.

References

1. Lambert, P. and Dykes, K. (1994) The Investigation and Repair of the Queensway Tunnel Road Deck. *SCA Seminar on Tunnel Construction and Repair with Sprayed Concrete*, 28 September, Loughborough University.
2. Hayward, D. (1995) Approach Shot. *New Civil Engineer*, 23 March, pp 26-27.
3. Austin, S.A. (1994) Introduction to sprayed concrete. *Sprayed Concrete: Properties, Design and Application*, (ed. S.A.Austin and P.J.Robins), Whittles Publishing, Caithness, p5.
4. Seymour, J. and Turner, N.J. (1995) Wet Process Sprayed Concrete Technology for Repair - Repair Scenarios. *Internal Report*, Department of Civil and Building Engineering, Loughborough University of Technology, U.K, November.
5. Turner, N.J. and Seymour, J. (1995) Wet Process Sprayed Concrete Technology for Repair - Mix Design. *Internal Report*, Department of Civil and Building Engineering, Loughborough University of Technology, U.K, December.
6. Tattersall, G.H. (1973) The principles of measurement of workability of fresh concrete and a proposed simple two point test. Fresh Concrete: Importnat Properties and Their Measurement, Proc. RILEM Seminar, 22-24 March, Leeds University, Vol.1, p2.2-1 to 2.3-33.
7. European Federation of National Association of Specialist Contractors and Materials Suppliers to the Construction Industry. (1993) *Specification for Sprayed Concrete (final draft)*, EFNARC, Aldershot, UK.
8. Beaupré, D. (1994) Rheology of high performance shotcrete. *PhD Thesis*, University of Colombia, Canada.

18 WET SPRAYED MORTARS REINFORCED WITH FLEXIBLE METALLIC FIBRES FOR RENOVATION: BASIC REQUIREMENTS AND FULL-SCALE EXPERIMENTATION

J.M. Boucheret
Fibraflex-Seva, Chalon-sur-Saone, France

Abstract
The renovation of brickwork and concrete structures is a necessity, as it means maintaining by adding value using efficient techniques and durable products. This paper reviews the latest developments of wet sprayed mortars reinforced with Fibraflex amorphous metallic fibres. Based on the designer's, the dry-packed mortar producer's, the applicator's, the client's point of view, we split up the issue in separate factors such as workability of the fibrous material, durability, mechanical performance. The second part of the study is a full-scale experimentation : renovation of ovoïd-shaped precast-elements, used in combined sanitary and storm systems. A crushing device was designed and results show that amorphous fibres advantageously replace traditional wire mesh.
Keywords : amorphous metallic fibres, durability, fibre reinforcement, mix design, ovoïd-shaped sewage systems, post-crack behaviour, renovation, wet sprayed mortars, workability.

1 Introduction

Nowadays, the renovation of a concrete or a brickwork structure means conserving a useful and essential patrimony with efficient techniques and long-lasting products. The structures concerned are mostly those built since the end of the 19th century, such as tunnels, bridges, dams, piers, water towers and in the present article mansized sewers. Sprayed concrete is being extensively used for underground structures and tunnel linings for both construction and renovation purposes [1].

Pont-à-Mousson, the French manufacturer of cast-iron pipes developed in the early eighties, Fibraflex, the amorphous metallic fibre, which has since been employed as a reinforcement medium for sprayed concrete and mortars [2], [3].

2 Wet sprayed repair mortars reinforced with amorphous metallic fibres

Renovation with sprayed mortars requires the combination of three attributes :
1. workability and ease of placing
2. durability
3. mechanical performance

Designing an efficient renovation mortar which complies with the above has required the experience of all the parties : the concrete producer (specific process), the applicator (equipement and site organization), design engineers and the owner.

2.1 Workability and ease of placing
Two different techniques are employed for sprayed concrete and mortars :

• the dry process, the oldest method, generally applies to mortar and concrete. Large equipment can be used including machine operated arm, for tunnels, stabilized slopes or foundation pits. The dry process generates dust, and loss of material due to rebound and particles flying off has a significant effect on the job-site efficiency; it is therefore being abandonned for the wet process as regards small-size underground structures like sewers, aqueducts, water galleries, tunnels or inspection galleries.
• the wet process is more recent as it emerged from the crossroads of building and civil engineering techniques, while admixtures technology was improving. It is a simple and clean way to spray mortars. Small size spraying equipment is chosen because easy to maintain and convenient to use when space is limited.

Amorphous metallic fibres geometry is a thin ribbon 20 mm long, 1.6 mm wide and 29 μm thick*. These fibres are very flexible and can easily be incorporated into the mortar, supplied in paper-bags, big-bags, silos or as ready-mixed concrete by the producer. The amount of fibres studied was 35 kg/m^3 or 0.5% vol. Mixing and pumping capacities are not altered, so standard machines (cork-screw or piston type) are still suitable [4]. Pumping distance is over 50 m, with 35 and 50 mm pipes.

2.2 Durability of fibrous mortars
Durability means choosing adequate cement and admixtures, together with corrosion resistant products. In this respect, amorphous metallic fibres (FeCr)$_{80}$ (P,C,Si)$_{20}$ happen to be extremely corrosion resistant under chloride, acid (HCl, H$_2$SO$_4$) or alkaline attack and in cementitious matrixes [5]. For this reason, Fibraflex was chosen by the French Nuclear Authorities Andra and Cogéma for the manufacture of fibre reinforced concrete containers used in low-level radioactive waste disposal. These containers are certified for a life of 300 years [6].

The corrosion resistance is also necessary in waste water systems. Moreover, the observation of cores bored out of the first renovated sewers (from 1986 onwards) should confirm the outstanding behaviour of Fibraflex in chemically aggressive environment [2], [4].

---

* Fibraflex FF20L6 : 150 000 fibres/kg, specific surface : 9.6 m^2/kg,
  tensile strength : approx. 2000 MPa.

2.3 Mechanical performance
The performance of the system is studied via three separate fields :
1. mechanical performance of the fibrous material itself
2. behaviour of the repair layer onto its base under load
3. overall behaviour of the repaired structure under load

Point 1. is treated in [7] and some of these results are described in Chapter 3.2 - Material employed. This mechanical characterization has been carried out by Pont-à-Mousson Laboratory, based on former experience of fibre reinforced concrete [8].
Point 2. is currently being studied by H.Chausson, Fibraflex engineer, as part of her PhD research at UPS Toulouse. Experimentation and finite element computation show that cracking of the repair layer and debonding are linked together, one of the key factors being the post-crack residual tensile strength of the fibrous material [9].
Point 3 is treated in Chapter 3.3 - Experimentation.

2.4 Basic requirements
As a conclusion of Chapter 2., the main issue has been split up into various key factors, each of these being investigated separately. Table 1. summarizes the influence of each step in view of designing an efficient wet sprayed repair mortar.

Table 1. Requirements vs. influential factors

	workability and ease of placing	durability	mechanical performance
mix design	X	X	X
performance of material		X	X
amount of fibres	X		X
chemical resistance of fibres		X	
flexibility of fibres	X		
behaviour of repair layer on base		X	X
overall behaviour of the structure		X	X

3 Full-scale experimentation

3.1 History
This experimentation concerns the renovation of man-size sewers which collect waste and/or rain water. Some of these were installed as early as at the Roman Empire era, but most of them were built during the massive urbanization and industrialization of major European cities at the end of last century [10]. Paris and its suburbs has the largest network of waste and rainwater collection in France ; consequently, renovation techniques have been experimented for a long time in this area.
Le Conseil Général du Val-de-Marne together with Abrotec implemented in 1985 an experimental system in order to assess the performance of sprayed concrete and mortars for the renovation of man-sized ovoïd-shaped sewers [4].
A precast element is sawn along one of its sides to simulate a crack in the substrate. A layer of mortar is sprayed into the interior by an experienced applicator. A crushing device is employed so that displacements are monitored as load increases.

3.2 Material employed

3.2.1 Design method

Sprayable mortars reinforced with amorphous metallic fibres are supplied as dry-packed products, in bags, big-bags or silos. Products available on the market are Emaco S170 CFR made by MAC (MBT Italy) or PSM-F 35 made by La Pierre Liquide (France).

The latter product was designed specifically for this kind of renovation. This mortar is based on CPA-CEM I-52.5 cement, 0/4 mm sand, limestone filler and 35 kg/m³ of Fibraflex FF20L6 fibres.

As mentionned in 2.4, the following key factors have been studied as a single issue in the laboratory : geometry and amount of fibres, mix design, workability and flexural performance of the mortar. The best formula was then chosen (workability and flexural performance) and the mortar was sprayed into panels in order to saw out test pieces and assess the behaviour under third-point load and centre point load on square plate, as well as compressive strength and direct tensile strength [7].

3.2.2 Compressive strength
The compressive strength was 40 MPa at 28 days.

3.2.3 Flexural strength

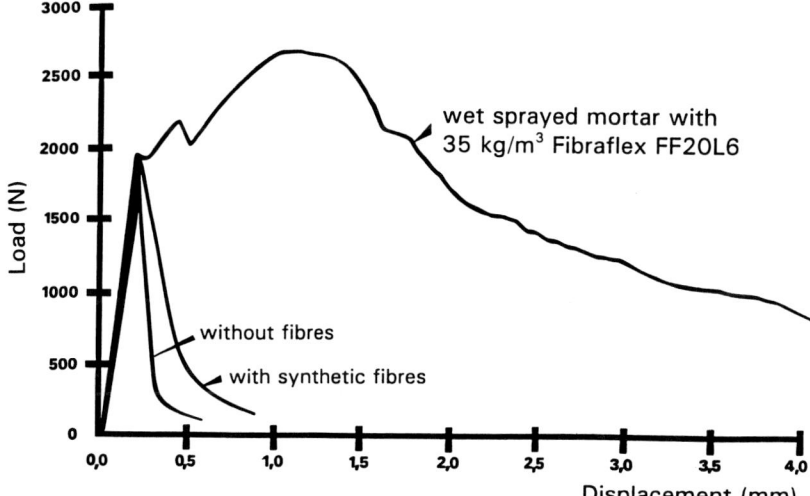

Fig. 1. Third point flexural test. Typical behaviour.
Specimen dimensions : 350 x 80 x 40 mm, span 300 mm.

According to ASTM C 1018, toughness indices are the following :
I5 = 5.35
I10 = 12.0
I20 = 23.3

3.2.4 Energy absorption capacity

The test plate (60 x 60 x 10 cm) is supported on its four edges and a centre point load is applied through a contact surface of 10 x 10 cm.

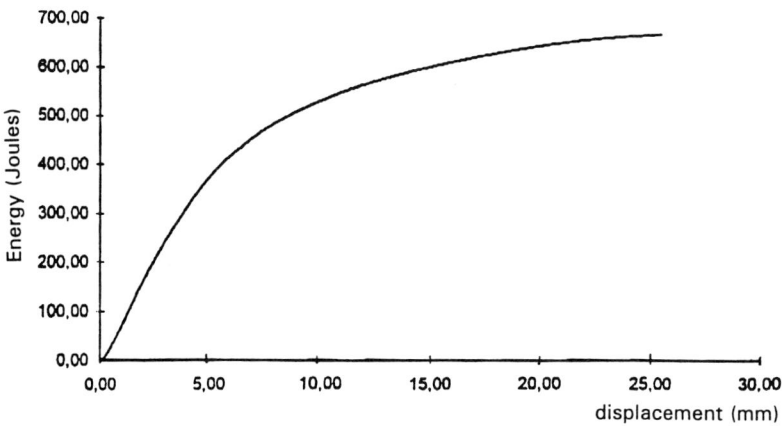

Fig. 2. Plate test. Typical behaviour. Maximum load : 95 kN.

3.2.5 Direct tensile strength.

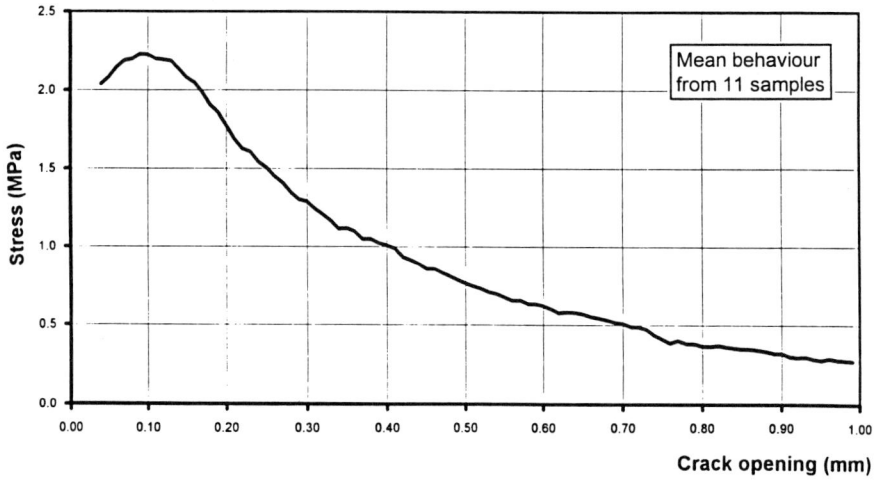

Fig. 3. Mean post-peak tensile behaviour.
Notched specimens : diametre 100 mm, height 100 mm, depth 15 mm.
AFREM procedures [11].

3.3 Experimentation

A previous study was based on shotcrete (dry-process) [12]; here the wet sprayed mortars are being tested. Spraying machines were Putzmeister P11, P13 and P38.

The tests were carried out on different pieces :
- a sawn piece (simulating a cracked structure);
- a new one (as if the sewer was being demolished and rebuilt).

Two renovation techniques using wet sprayed mortars :
- the traditional method : a wire mesh is installed (P400, 2.475 kg/m^3) and 7 cm of mortar is applied. 7 cm to 10 cm is the minimum thickness to be applied as rebars must be properly encased in concrete in order to reduce the risk of corrosion, especially in the chemically aggressive environment of waste water;
- the second one is a layer of 5 cm of Fibraflex mortar, as described in 3.2.1.

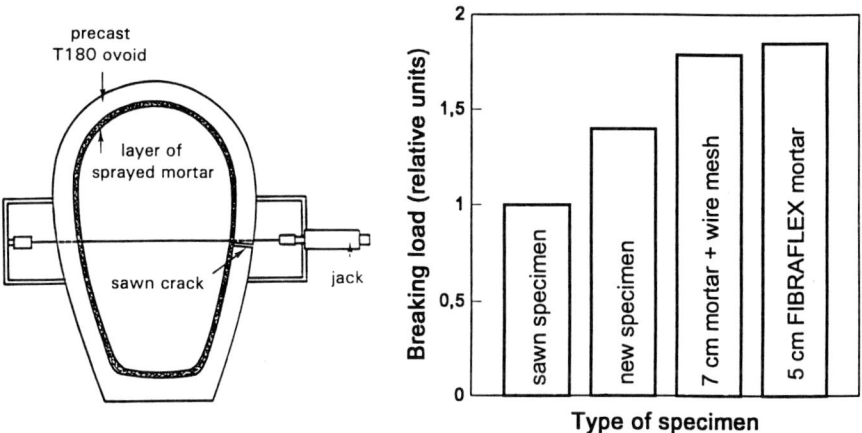

Fig. 4. Experimental apparatus   Fig. 5. Crushing load at breaking point

These results show that Fibraflex mortars reinforce the structure well above the strength of the unsawn piece and offer similar strength as the steel mesh.

4 Conclusions

This experimentation provides useful comparative data on the performance of different renovation techniques. Amorphous metallic fibre mortars are desirable when working in cramped conditions, as the equipment is small and placing is easy. Economy is significant since placing the mesh is no longer required. Savings to expect can be 10% of the cost of the spraying job and the duration of the total job 12% less.

Post-crack residual strength (fig. 1, 2 and 3) has a positive effect on the overall behaviour of the structure under load. Future developments will be focused on finite element computation, based on the direct tensile mechanical behaviour of the material, in order to produce design guidelines for renovation purposes.

5 References

1. Resse, C., Venuat, M. (1981) Projection des mortiers, bétons et platres. Techniques et applications au BTP, Paris.

2. Burgun, D., de Guillebon, B. (1987) Béton projeté renforcé de fibres de fonte. *Revue Techniques Sciences Méthodes*, AGHTM, Paris.

3. Boucheret, J.M. (1994) Concrete and mortars reinforced with FIBRAFLEX amorphous metallic fibres. Workshop on fibre reinforced cement and concrete. University of Sheffield, Sheffield.

4. Resse, C. (1994) Emploi des fibres dans les bétons projetés. Premier Colloque Francophone sur les Bétons de Fibres Métalliques. Béthune.

5. de Guillebon, B. (1987) Résistance à la corrosion de la fibre de fonte. Centre de Recherches de Pont-A-Mousson.

6. Pech, R., Schacher, B., Verdier, A. (1992) Fibre reinforced concrete containers : from concept to fabrication. *Internal Symposium on Fibre reinforced Cement and Concrete*, Fourth RILEM, Sheffield.

7. Peiffer, G. (1996) to be published. Deuxième Colloque Francophone sur les Bétons de Fibres Métalliques. Toulouse.

8. Peiffer, G. (1994) Mix design and mechanical properties of high performance concretes reinforced with amorphous metallic fibres. *1994 ConChem Conference*. Verlag für chemische Industrie, Augsburg.

9. Granju, J.L., Chausson, H. (1995) Serviceability of fiber reinforced thin overlays : relation between cracking and debonding. *1995 ConChem Conference*. Verlag für chemische Industrie, Augsburg.

10. Boucheret, J.M. (1995) Réhabilitation par projection de bétons et mortiers fibrés. Choisir et mettre en oeuvre une technique de réhabilitation. Office International de l'eau. Limoges.

11. Méthode de dimensionnement. Essais de caractérisation et de contrôle. Séminaire Les Bétons de Fibres Métalliques. AFREM, Saint Rémy Les Chevreuses.

12. Schacher, B. (1989) Béton projeté renforcé de fibres de fonte FIBRAFLEX pour réseaux visitables. *Revue Techniques Sciences Méthodes*, AGHTM, Paris.

19 SPRAYED MORTAR REPAIRS TO HIGHWAY STRUCTURES

R.B. Dobson
Frank Graham Consulting Engineers, Wakefield, UK

Abstract
The maintenance of highway structures requires engineers to overcome very demanding constraints. Whenever possible traffic flows must be maintained, and unavoidable delays reduced to minimum. Access for repair and refurbishment can often be difficult and a high quality of work must be produced in the shortest possible time. The paper presents a number of case histories which demonstrate the role that sprayed mortar can play in achieving these objectives.
Keywords: Sprayed mortar, highway structures, maintenance, refurbishment, repair

1 Introduction

The importance of achieving the full design life of highway structures through effective maintenance and refurbishment is recognised around the world. In the United Kingdom, the Highways Agency has implemented a 15 year programme for the rehabilitation of Motorway and Trunk Road bridges [1]. However, traffic volumes on the Motorway and Trunk Road network are high and continue to rise. Reducing the impact of this work presents a continual challenge to engineers.

Measures taken to reduce the delay to the travelling public, include working in off peak periods, quite often during the night , the use of lane rental contracts and incorporating the work in major carriageway reconstruction contracts. The effectiveness of these measures is enhanced by the use of well planned and efficient repair techniques.

The paper will draw on the experience gained by Frank Graham, on the West Yorkshire Term Maintenance Commission , in the development of repair techniques using sprayed mortar. This will be demonstrated by the presentation of four case histories:

1. Altofts River Bridge - the replacement of chloride contaminated concrete in the reinforced concrete deck cantilevers.
2. River Aire Bridge - a parapet anchorage replacement scheme.
3. Motorway overbridge deflection points - repairs to numerous small areas of concrete in the soffit of prestressed concrete bridge beams.
4. Oxford Road Bridge - abutment repairs.

2 The West Yorkshire Term Maintenance Commission

Frank Graham managed the maintenance of the Motorway Network and part of the Trunk Road Network in West Yorkshire for the Highways Agency from April 1986 to March 1996. The total length of the highway network is 175km and includes the M1 and M62 motorways and A1 trunk road.

The majority of the 340 structures on the network are of modern reinforced concrete, prestressed concrete and composite steel/concrete construction Structural types range from small diameter culverts to multi-span viaducts.

Routine maintenance operations are generally carried out by a Term Contractor appointed for a fixed period of time. Larger items of bridgeworks are let on a competitive tender basis, both as part of major highway maintenance contracts and as independent bridgework contracts.

Traffic volumes on the network are high for example on the dual carriageway A1 (north of the M62), the annual average daily traffic (aadt) is 54,500 vehicles per day of which 29% are HGV's (1994). The most heavily trafficked motorway section is the M62 between junctions 26 and 27 with an aadt of 102,500 vehicles (21% HGV) (1994)

3 Case Histories

3.1 Altofts River Bridge

The bridge is a three span continuous composite structure carrying the M62 motorway over the River Calder. The bridge superstructure consists of four steel box beams with regularly spaced transverse steel beams supporting a 225mm diameter reinforced concrete slab. The central span is 54m in length and the side spans 27m.

A longitudinal joint separates the two halves of the superstructure (Fig. 1.). The light form of construction of the bridge deck produces significant differential movement at the central joint. Attempts to seal the joint had proved unsuccessful and heavy leakage of water containing de-icing salts had occurred

A principal inspection of the bridge revealed severe chloride contamination of the soffit of the cantilevered deck slabs adjacent to the longitudinal joint. Large areas of concrete were found to be delaminated and spalling. A comprehensive concrete investigation was undertaken, which indicated high values of chloride content well beyond the level of the reinforcement.

Sprayed mortar repairs to highway structures **175**

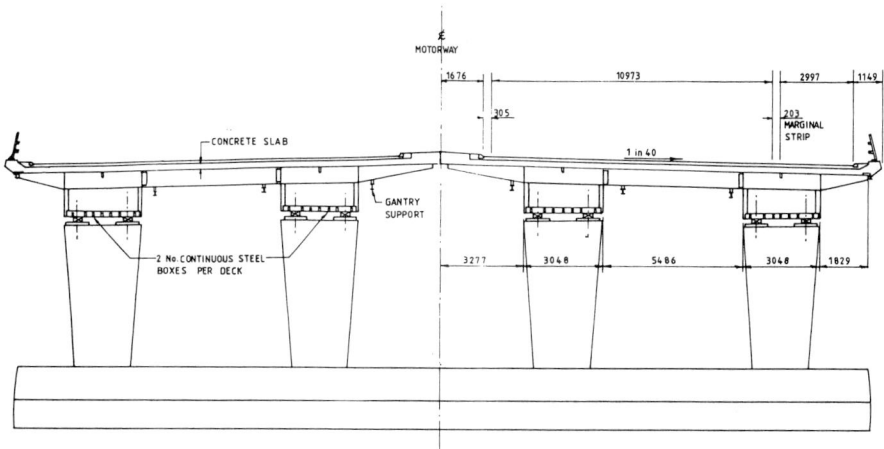

Fig. 1. Altofts River Bridge - typical cross section

Fig. 2 shows the original detail of the deck cantilevers. The complete replacement of the deck cantilevers was considered initially. However, lane closures would have been required on the motorway for a considerable period of time.

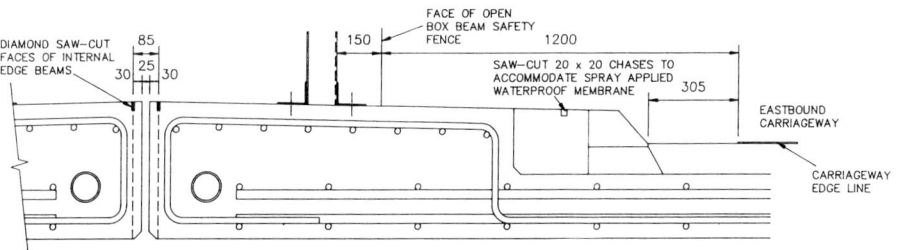

Fig. 2. Altofts River Bridge -original detail of deck cantilever

A repair scheme was finally adopted using sprayed mortar to replace the defective concrete. The majority of the work would be carried out from below the bridge, removing the need for permanent lane closures. Another consideration, was that traffic induced vibration would occur in the reinforcement during placing of the replacement material. Sprayed mortar had the added benefits of a high degree of compaction, good adhesion and early bond strength.

A suspended scaffold system, was erected below the bridge to provide a solid working platform. All materials had to be brought to a single access point on the motorway hardshoulder and carried along the scaffold. Temporary lane closures were required for some operations, such as saw cutting to widen the existing joint. This work was undertaken at night/weekends to limit delays to traffic. The detail of the repair is shown in Fig. 3.

The existing defective concrete was removed from the soffit of the cantilever using high pressure water jetting. Concrete was removed to the depth of chloride contamination, as determined by the concrete investigation. The areas to be repaired were divided into panels, typically 6m by 2m. The sequence of removal of each panel of concrete was designed to maintain the structural integrity of the deck. Restrictions were also placed on the areas of concrete that could be removed, before the new areas of mortar had gained strength.

An epoxy modified cementitious chloride barrier was applied to the concrete prior to the application of the proprietary spray applied mortar. The average thickness of mortar was 100mm which was applied in a single layer. Removal of rebound material proved to be a difficult operation due to the restricted access.

A plastic fascia panel was attached to the edge of the of the deck, using stainless steel fixings. This acted as formwork for the mortar and also provided a drip feature, to reduce the effect of any future leakage.

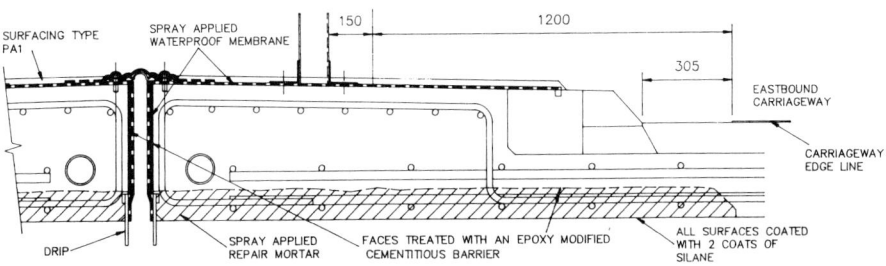

Fig. 3. Altofts River Bridge - deck cantilever repair detail

The mortar was trowelled to a U3 finish, which produced a smooth dense high quality surface. Cores were taken from the mortar, and strengths of over 36N/mm² were obtained after three days. The cores also confirmed that a good bond had been achieved between the mortar and the reinforcement.

The vertical surfaces of the joint were finally sprayed with a waterproofing membrane and a rubber seal clamped over the widened gap between the decks. This detail is designed to accommodate the differential movement.

The scheme was let as a conventional contract and was completed within the 20 week contract period.

3.2 River Aire Bridge

The River Aire Bridge is a major three span balanced cantilever structure, carrying the A1 Trunk Road over the River Aire and B6136 at Ferrybridge in West Yorkshire. The bridge superstructure consists of four post tensioned variable depth concrete box sections, total span approximately 200m.

Replacement of the existing steel P1 parapet with a new aluminium P2 parapet involved extensive strengthening work to the parapet cantilevers. The existing reinforcement in the cantilever was not sufficient to meet the current loading standards for the parapet anchorages.

Local strengthening was necessary at 140 No. post locations. This required the removal of an area of existing concrete typically 0.85m wide by 1.20m long and 0.16m deep from the cantilever upstand, and a 0.16m by 0.15m section from the parapet downstand see Fig.4. Additional reinforcement was to be fixed in each location prior to replacement of the existing concrete and the installation of resin parapet anchors.

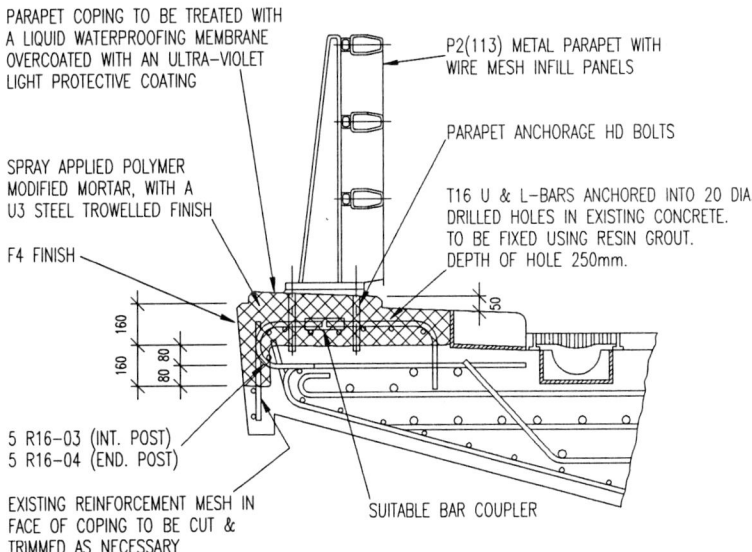

Fig. 4. River Aire Bridge - parapet anchorage strengthening detail

The work formed part of a much larger maintenance contract, the A1 River Aire to A63(T) Reconstruction Contract. In order to minimise the length of the contract and consequent delay to traffic the contract was to be let on a lane rental basis. The bridgework represented a major element of the contract and any delay in the completion of each individual area would be magnified many times. Traffic would continue to use the bridge during the works and the induced vibration in the structure was again an important consideration. For these reasons sprayed mortar was adopted as the replacement material.

Hydrodemolition was preferred to pneumatic techniques for the removal of the existing concrete in order to avoid,
1. Damaging post tensioning ducts that were shown in the cantilevers on the as-built drawings.
2. Microcracking in the surrounding concrete.

An interesting feature of this work was the use of robotic hydrodemolition, operating at a pressure of 12000 to 14000 psi (83 to 97 N/mm^2), to achieve precise excavation of the main parapet cantilever concrete. The downstand was broken out using a hand lance at 16000 psi (111 N/mm^2).

The application of the sprayed mortar reinstatement material was very successful, producing a trowelled finish which blended well with the adjacent concrete.

3.3 Motorway Overbridge Deflection Points

Pre-tensioned prestressed bridge beams are a common feature of many of the overbridges on the West Yorkshire motorway network. Variations in the profile of the strands were produced by the use of steel "trees" positioned at predetermined locations in the mould, see Fig. 5. The base of the "tree" was positioned at approximately 50mm from the soffit of the beam and tied by a strand through a recess in the beam.

In some of the beams, the strand was burnt off after casting, and the recess filled with a proprietary quick setting mortar. A number of the bridges were subsequently found to have defects associated with the deflection points as a result of the high chloride content of the mortar.

Investigation of a representative sample of affected beams revealed that small amounts of corrosion had started to develop in the strand at the deflection points. Therefore, a comprehensive programme of remedial work was implemented in order to prevent serious damage.

The remedial works were undertaken by the Term Contractor as part of the West Yorkshire Term Maintenance Contract. Payment was made in accordance with a schedule of rates which formed part of the contract.

Although the areas to be repaired were small, generally 200mm x 300mm, their location, over the full width of the carriageway, presented difficult logistical problems. The work had to be restricted to off peak periods, because of the number of traffic lanes required for each series of repairs.

The use of hand applied repair materials was considered, but this would have required the application and curing of a number of individual layers. Significant savings in cost and occupation of the carriageway were achieved by using spray applied mortar.

The repair technique was developed on a canal bridge, located below the motorway, where access was relatively straightforward.

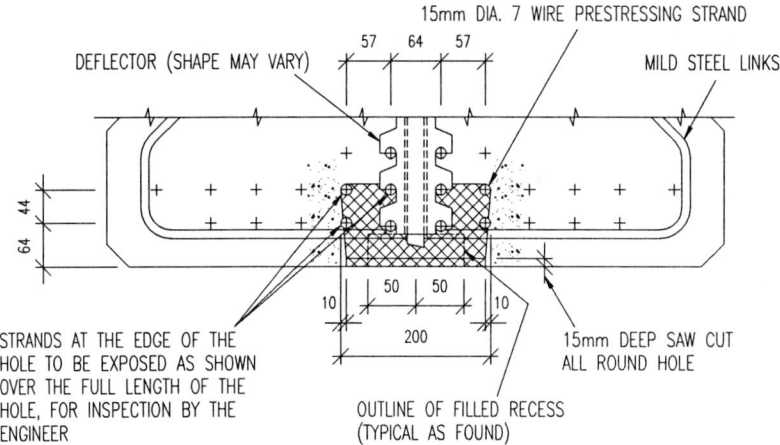

Fig. 5. Section through a prestressed beam showing the steel tree and repair area

The results of a previous concrete investigation programme were used to define the depth of the repair. High pressure water jetting was used to remove the concrete. The edges of the repair area were first sawcut to provide a clean working edge.

Dry process sprayed mortar was selected because of the relatively small areas of repair. The total volume of repair was maximised whenever possible by first breaking out a line of deflection points across the width of the bridge, i.e. within one lane of traffic management. Even more efficient use of materials could be achieved by combining repairs on adjacent bridges.

Protection of passing motorists from spray and debris was an important consideration. Extensive screening was avoided by using a scaffold tower shrouded with polythene sheeting. The tower was jacked up to the underside of each beam so that any spray/debris fell on to the carriageway below the tower.

A total of 2253 deflection points have been treated using this technique, on 53 bridges. The repairs were carried out in 1992/1993 and have continued to perform satisfactorily.

3.4 Oxford Road Bridge Abutment Repairs

Oxford Road bridge is a three span reinforced concrete structure carrying the A651 over the M62. A concrete investigation into the south abutment revealed excessive chloride contamination, principally as a result of traffic spray. Typical chloride contents were of the order of 1.5%, by weight of cement, with maximum values of 4.1%. The area of contamination was so extensive that the decision was taken to repair the complete face of the abutment.

The hardshoulder and verge adjacent to the abutment were separated from the main carriageway by Temporary Vertical Concrete Barriers[2]. This enabled three lanes to remain open on this heavily trafficked section of the motorway. Due to the proximity of the abutment to the carriageway, a protective screen was erected around the works. The work was carried out in November and space heaters were used to maintain a reasonable temperature.

The face of the abutment was removed to 25mm behind the level of reinforcement, a depth of 100 to 120mm. Again, high pressure water jetting was the preferred method for concrete removal. Hand held lances were used at pressures of up to 30,000 psi (207 N/mm^2). An added benefit of water jetting is that it cleans the reinforcement. However, in this case the reinforcement was so heavily corroded that grit blasting to Swedish Standard 3 was also specified.

A proprietary, dry process sprayed applied mortar was specified with a specified grade of 40 N/mm^2 , a water /cement ratio between 0.35 and 0.45 and a minimum cement content of 400 kg/m^3 . Typical wastage was around 25%.

The finished repair was given a floated finish and various features in the original face, fillets and rebates, were successfully repeated. Cores were taken from test panels (100mm dia) and the completed surface (50mm dia). Values of 23 N/mm^2 to 30 N/mm^2 were recorded after 24 hours.

The combination of cold weather and general air turbulence around the repair area, is thought to have lead to some cracking in the face of the repair material, despite careful curing. Following investigation into the cause of the cracking, a two part sealing liquid was applied to the whole of the face.

4 Conclusion

The case histories demonstrate that sprayed mortar techniques can be used to provide variation in the approach to maintenance repairs. Particularly where demanding constraints make conventional methods less attractive.

5 References

1. Department of Transport. (April 1993) *TRMM 2/90, 15 year rehabilitation programme.*
2. DOT/TRL. (1990) *Drawings 1040.66 Series. Temporary vertical concrete safety barriers.*

6 Acknowledgements

The author would like to thank Kevin Lasbury, Director of the Highways Agency's Northern Network Management Division for his permission to publish this paper. Any views expressed in the paper are those of the author and do not necessarily represent those of the Highways Agency. Thanks are also due to many colleagues for their assistance, in particular Mark Christian and Richard Feast.

20 AESTHETIC FLATWORK REHABILITATION

J.H. Ford

Symons Corporation, Des Plaines, USA

1 Abstract

There are millions of exterior concrete slabs, driveways and walkways in the USA today that are scaled, severely discoloured or otherwise aesthetically displeasing. Repair and re-habitation of this concrete affords an excellent opportunity for the sprayed concrete industry. The use of aesthetically pleasing sprayed concrete and creative use of both colour and texture is a feasible, practical and an expeditious method of attaining an aesthetically pleasing and durable concrete. Rehabilitation when performed in conjunction with good concrete repair techniques, can achieve desirable, aesthetically pleasing results.

Due to the constraints of this paper, I will only discuss a direct contractual arrangement between contractor and owner, without the use of an architect or engineer.
Keywords air-entraining, scaling, curing, polymer-modified, shrinkage, creep delamination.

2 Causes

The interstate highway system in the USA, used 6% +/- 1-1/2 %[i] of air-entrainment agents in the paving concrete, which has proven to be very durable. Water reducing admixtures were also used in the interstate system[1], but there was minimal use of high range water reducing admixtures (HRWRA). The mix designs for the interstate system started with a high fineness modulus[ii] for the fine aggregate and the maximum coarse aggregate size was rarely smaller than 1-1/2"[iii]. With the high fineness modulus of the fine aggregate, a larger maximum aggregate, only small gaps in the gradations of either the fine or coarse aggregates and minimal use of high range water reducers, this percentage of air entrainment did not produce the stickiness and rubberiness often associated with mixes today that have low fineness moduli of fine aggregate, small coarse aggregate, gap gradation of both fine and coarse aggregate, high air entraining and high percentages of HRWRA.

The majority of all the interstate system concrete paving was: mixed, placed, consolidated and struck-off by paving machines, then straight-edged manually, and finally bagged and cured by machine; no steel trowelling was used or required.

For the last twenty years, a significant amount of the residential and commercial exterior concrete slabs placed on the ground used ready mixed concrete, with high air entrainement, lower sand fineness, smaller aggregate.

These mixes can be sticky and can be floated easily with wooden floats that keep the surface open. This stickiness forces contractors to use magnesium/aluminum float in lieu of the wood floats and/or steel trowels (Fresno's, a long handled steel trowel with rounded corners) that induce fine migration to the concrete surface, in lieu of the wood float.

The excess mix water that comes to the surface of the concrete (bleeding) is minimal when the concrete is air entrained. The ability to determine precisely when the bleeding stops requires training and development of this skill by the finisher. If steel trowelling takes place before the bleed water to rise to the surface and evaporate, premature sealing takes place. In this situation, when the bleed water rises, it collects at the bottom of this seal layer producing a weak, poorly bonded layer just below the fi^{iv}nished surface. In time, the thin finish-layer delaminates and establishes scaled concrete. The concrete below this sealed layer is well cured as the bleed water acts as the curing water. The ultimate delamination of the sealed layer does not occur for a few days. With the freeze/thaw resistance produced by the air-entrainment throughout the concrete, but below the sealed layer, these scaled slabs are often durable, though aesthetically unpleasant. Many exterior residential and commercial slabs placed in the last twenty years have been steel trowelled prior to the cessation of the bleeding, producing scaled slabs.

3. Slab Scaling:

Poor workmanship often produces scaling due to over-working or early working of the concrete surface with steel trowels, or excessive pitch on the float, poor or non-existing isolation joints that produce faulting or partial isolation joints and too shallow or poorly placed contraction joints that produce unintended cracks.

Impurities in either the fine or coarse aggregate can cause poor cohesion between cement paste and aggregates resulting in scaling or ravelling of the concrete.

Poor or no consolidation of exterior slab concrete plus hard trowelling can also produce scaling.

Delayed or no curing often results in a significantly weakened layer at the concrete surface, often with plastic shrinkage cracking.

Concrete that experiences no temperature control can result in questionable durability with isolated ravelling in spots.

4 Effects

The effects of extensive slab scaling, discolorations, low concrete densities, plastic shrinkage cracking, incorrect or no joints and uncontrolled drying shrinkage cracking, has established an extensive need for an aesthetically acceptable means of rehabilitation.

5 Rehabilitation

Rehabilitation usually begins with forensics (visual observations).

Forensic observations, when recorded on a drawing, are extremely valuable in concrete repairs of this type. Horizontal work should include both a plan view and sections. Vertical work should include elevation views and sections. There also needs to be a listing of the problem areas such as cracks, joints and surface repairs. When the list is complete, it is also necessary to establish whether the intended repair is applicable to both the ACI Building Code (318) and Local Building Codes.

In any concrete repair, the more knowledgeable the repair contractor is of the causes of concrete problems, the better the chances for a long-term, durable repair.

Cracks should be designated as plastic shrinkage cracks, drying shrinkage cracks or moving cracks. There are three types of joints used in USA: Isolation (allow differential movement both horizontally and vertically), Contraction (allow differential horizontal movement only), and Control (joint is restrained both horizontally and vertically).

Forensic field testing can often satisfy the needs of most residential and commercial repair needs. Chain dragging has been used for years, but requires an operator with good ears, and detailing on a plan of the problem areas.

A logistical plan can now be established allowing the supplier of materials to properly place materials in pre-determined locations. This assures that the correct material is used and will be delivered and deposited efficiently. When a logistical plan is included, the productivity of the work will greatly improve.

6 Repairs

When a plan for the rehabilitation has been agreed upon, the contractor needs to produce a bill of materials for the project.

Horizontal repairs should be performed prior to surface preparations.

Plastic shrinkage cracks are usually very visible. A sealer/healer composed of high molecular weight methacrylicates[v] not only will seal plastic shrinkage cracks but will also bond them structurally. Broadcasting sand into the neat methacrylate will assist bonding of the polymer modified cementitious overlay.

Dormant drying shrinkage cracks should be sealed when the restraint that produced the unintended crack in the first place has been eliminated.

Moving cracks should be routed and finished similarly to joints. It is absolutely essential that the owner be fully aware of the final appearance of moving crack repairs as they are almost always not a function of the original design, nor can they be easily incorporated into the new design.

Existing joints should be detailed on your drawing as either: isolation, contraction or control. It may be necessary to make either a control or contraction joint into an isolation joint. Ravelled joint nosing must be repaired with a cementitious or epoxy material that satisfies the ultimate usage of the slab. The exposed width of all joints must be consistent, and the treatment procedures and materials should be uniform.

All repairs must be in accordance with existing building codes.

Vertical surface repairs also should be completed prior to surface preparations and are very similar to the repairs in horizontal concrete. However, exterior retaining walls should be checked for plumb and any other measures to ensure that they are still retaining their over-burden and have not started to tip over. If tipping has started, removal and replacement may be required.

Delaminations or other ravelled concrete must be saw cut to a reasonable depth and the bad concrete removed.

Surface offsets should be re-profiled with a polymer modified mortar, and the finished surface brushed to assist the bonding of the overlay.

7 Surface preparations

The first consideration is the method of surface preparation: mechanical or chemical. Mechanical surface preparation such as grit blasting, shot-tracing, bush-hammering, water blasting or a combination water and grit blasting, shot blasting, bush hammering, and water blasting or a combination water and grit blasting generally remove deleterious material, and prepares the surface for the sprayed concrete overlay.

Chemical surface preparation should only be considered when the concrete surface to be repaired is structurally sound. A caustic cleaner will remove hydro-carbon emissions as well as oil and gas spills; heavy oil penetration should be removed mechanically. Caustic cleaners can remove most vegetation pollen. Oxidants can remove vegetation stains and mildew. Acid cleaners should only be of a non-chloride type. Acid cleaners can remove cement dust and thin layers of laitance. Chemical removal is labour intensive, often not environmentally preferred, and causes clean-up and disposal problems, but can be much less expensive on small projects.

Sealing of joints and moving cracks is best performed prior to the first spray application. A masking tape placed over the sealant, and not removed until the spraying has been completed, will produce a crisp, workman-like finished product.

8 Selecting Spray Materials

Thin sprayed decorative overlays are often polymer modified cementitious products (PMC).

The most widely used polymers for polymer modified cementitious mortars are:
Poly-Vinyl Acetate (PVA)
Styrene-Butadiene Resin (SBR)
Poly-Acrylic Esters (acrylics)
Styrene Acrylic (SA)

PVAs are re-emulsifiable both before and after the overlay has cured, and should never be used in damp conditions SAs are also re-emulsifiable, but not as severely as PVAs. SBR has the best properties overall for this type of spray application, but some blends can cause discolouring of light colours with time Acrylics are also very effective and are less susceptible to discoloration problems than SBR, but are more expensive. Therefore, the best polymer modification materials for sprayed cement for rehabilitation are acrylics or SBR.

A mix water produced with 10--15 % SBR by weight is the best polymer modification when used with a proprietary blend of sand, cements, pigments and admixtures for polymer modified sprayed mortars. Re-dispersed powdered latex is also used as part of the powder mixture, but some of the physical properties of the finished overlay can decline as much as 15 %.

Most repair mortars use white cement and white sands, and pigments.

9 Quality Control Requirements

Repair materials require far greater quality control procedures that similar materials for new construction. A manufacturer should periodically test and closely monitor all spray applied repair materials for:
Restraint to Volume Changes ASTM-C-1090
Modulus of Elasticity ASTM-C-469
Tensile Strength ASTM-D-638M
Bond Strength[vi]
Slant-shear ASTM-C-881
Linear Shrinkage & Thermal Expansion ASTM-C-531
Chemical Resistance ASTM-D-2299

10 Spray Equipment

The type of equipment used is low velocity (low pressure) and uses either a hopper-gun type or a slurry pump(mixer/sprayer combinations).

The hopper-gun type is lightweight and very portable and usually outputs at 0.2 m^3/min.--0.3 m^3/min.(7--10 cfm) at 0.2 MPa (30 psi), requires a separate air compressor, and must operate at low pressures. It is essential to have good control valves with gauges to ensure accuracy and consistency of pressure. The small hopper often requires repetitive mixing of spray materials. Tight quality control of the mix ratio (water to powder) is essential to assure colour consistency.

The slurry pumps are much more accurate at low pressures [0.2 MPa (30 psi)], are high performance [1.5 m^3/min. (50 cfm)], and are more efficient for larger projects. The hopper size is rather large requiring less batches, which affords greater consistency of colour.

11 Applications

Sprayed Base Coat

A workable consistency requires very minimal mixing, at low speeds. The latex admixture must be proportioned in advance of the start of mixing and labelled as ready to use. Always add the powder to the blended mix water and mix. Add the mixed mortar to the sprayer hopper and spray the mortar. For differential elevations greater than 3mm (1/8") the base coat should be trowelled

As in any aesthetically pleasing concrete, consistency of the material is essential, but more importantly, there must be no variation in the water/powder ratio. Minor variations in the water/powder ratio are acceptable in the first coat of a two coat job. Some suppliers can supply pigmented stains to offset colour variations. This procedure requires wet samples and an excellent eye of the workman. Re-tempering (softening a hardened mortar with water) any mortar is not a good practice.

Polymer modified cementitious mortars are self curing in calm air. If wind protection is not provided, intense plastic shrinkage cracking will result.

Membrane-forming curing compounds are not required for either coat if the surface is protected from wind. Water curing with sprinkling and mats can distort the colour of the repair. The overlay can be destroyed if watered when the overlay is not fully set. The best solution to curing is to plan wind protection and apply the finish coat as soon as the base coat is dry to the touch.

Due to curing and temperature protection constraints, logistical planning of this type of work is essential. Logistical planning consists of a layout of materials, close to their final destination. Crew planning also needs to be established well in advance, designating the mixer, applicator or nozzleman, the equipment operator and finisher.

Sprayed Finish Coat

Multiple colours along with design templates can add aesthetically pleasing effects at reasonable cost without long term delays in using the rehabilitated concrete when applied as a finish coat.

Light trowelling of a freshly sprayed surface produces a variably smooth top surface with fissures (differential offsets) produced by the random pattern produced by spraying. This process also decreases the surface profile resulting in a smoother overall texture and is referred to as a travertine marble effect. A trowel with rounded ends (pool trowel) eliminates most trowel marks.

Very light sprays of differing colours can also produce pleasing affects.

UV resistant sealers should be used and should allow a small percentage of vapour transmission. Sealers must be UV resistant when SBR is the polymer modifier and should be either a matte, semi-gloss finish for exterior applications, or gloss sealers for interior applications. Polymerizing sealers, Silanes and Siloxanes can be used in areas of high salting.

13 Conclusions

In any repair/rehabilitation system, it is mandatory to determine the cause of the problem or failure in order to recommend a suitable repair or rehabilitation. There is a need for sprayed astheticially pleasing repair in much of the world, but successful sprayed repairs require the services of an applicator who is knowledgeable of the causes of the problems in existing concrete and the effects, physical and chemical properties, limitations, features and benefits of the repair materials.

---

[i] US Highway Department "Manual of Instruction for Concrete Proportioning and Testing"
[ii] US Highway Specification Art. 703.01
[iii] US Highway Specification Art. 704.01
[iv]
[v] Symons S-70 or equal.
[vi] Kuhlmann, L.A.,"Test method for Measuring the Bond Strength of Latex-Modified Con August 1990.

21 SPRAYED CONCRETE REPAIRS – THEIR STRUCTURAL EFFECTIVENESS

G.C. Mays and R.A. Barnes
School of Engineering & Applied Science, Cranfield University,
Royal Military College of Science, Shrivenham, UK

Abstract
Spray applied polymer-modified cementitious based materials are now frequently used for the repair of deteriorating or damaged concrete structures. This paper addresses the structural effectiveness of such materials when used in large volume patch repairs. It describes the construction of 1:2.5 scale reinforced concrete frame structures, their repair under various load states and their load testing to failure. Comparisons of behaviour are made with theoretical predictions. The research showed that for the spray applied material investigated, the repaired section behaved in a similar way structurally as if unrepaired, thus giving some confidence in the use of such materials in structural situations.

Keywords: Cementitious materials, patch repairs, reinforced concrete, spray applied repairs, structural behaviour.

1 Introduction

The traditional materials used for the surface patch repair of reinforced concrete include hand applied polymer mortars and polymer-modified cementitious systems, and poured superfluid microconcretes [1]. More recently, there has been considerable interest in the use of spray techniques for the application of repair materials. This process can benefit considerably from the use of more sophisticated materials than simple sand-cement and accelerators. Many of the modern materials designed for spray application have been developed from the polymer-modified formulations originally developed for hand application. These provide more cohesive mixes which may be formulated with light-weight aggregates, thus permitting higher build. Trowel finishing is possible without damage or disturbance to the bond and, with appropriate formulation, mesh reinforcement for crack control may be avoided. These materials also display the same benefits of improved mechanical properties and better resistance to the penetration of chlorides and carbon dioxide as do their hand applied counterparts.

In larger volume patch repairs the repair material may be called upon to replace load bearing material and restore bond, in addition to the more usual aesthetic and durability objectives of concrete repair. A review of earlier work associated with the potential importance of property mismatch between patch repair materials and the reinforced concrete substrate has been given by the authors elsewhere [2]. More recently, Cairns [3] has reported on some alternative strategies for temporary support during the structural repair of reinforced concrete beams. In his work the repairs were effected by use of conventional concrete. Nevertheless, it is relevant to note that the ultimate strength of the beams repaired under load was not reduced as compared with control specimens, although there was some affect on ductility, deflection, crack widths and flexural stiffness. Further, Cairns concludes that there is unlikely to be significant benefits in postrepair performance as a result of relieving the structure of load during the repair process.

This paper describes the outcome of a programme of research aimed at investigating the structural effectiveness of large volume patch repairs to concrete structures. The programme has included a range of generic repair materials used in reinforced concrete sections subject to axial and flexural loads and has been previously reported on elsewhere [4,5]. However, the purpose of this paper will be to concentrate on the outcome of that part of the research which utilised a spray applied polymer-modified cementitious repair material, with the aim of providing designers with some confidence in the use of such materials in large volume structural patch repairs.

2 Model studies

Reinforced concrete H-frames were designed on a scale of 1:2.5 to represent a typical internal ground-floor frame in a reinforced concrete frame building or a bridge substructure. Typical design loads were scaled to provide loadings for the model beams and columns, a plane frame analysis conducted and reinforcement designed to BS 8110 [6]. Fig. 1 shows schematically the arrangement of an H-frame within the loading rig.

The repair location in the beam soffit was placed symmetrically about midspan and was of length 1300 mm, thus extending into the shear spans. The repair to the top of the beam was limited to 500 mm to allow for loading to be maintained on the beam during the repair process. The repairs on opposite faces of each column were 400 mm long and located centrally in the lower column legs. All the repairs extended across the full width of the section and were 50 mm deep, thus extending 20 mm behind the reinforcement.

The repair areas were prepared 28 days after casting of the H-frame concrete. The edges were saw cut and the concrete was then carefully broken out using a light demolition hammer. Grit blasting was employed to prepare the repair areas on the beam and one column, water jetting being used on the remaining column.

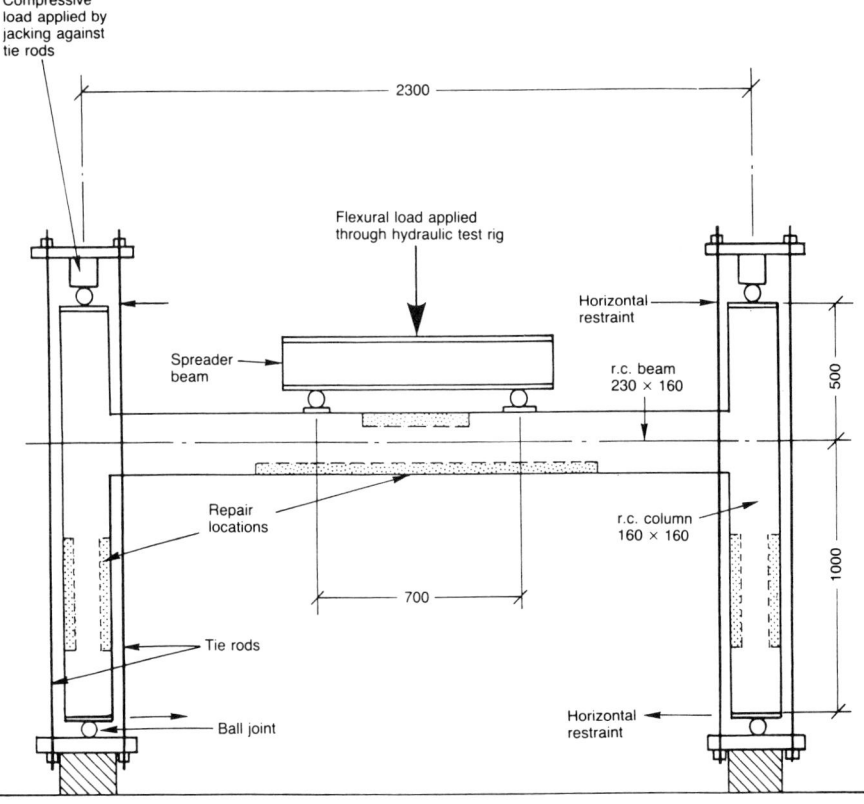

Fig. 1 Schematic arrangement of H-frame within loading rig.

Three H-frames were repaired with the spray applied material and these, together with one unrepaired control frame, were loaded as follows:

1. Frame 1, no load at repair (representing a propped structure).
2. Frame 2, equivalent of dead load at repair (column, 60 kN; beam, 15 kN).
3. Frame 3, equivalent of combined dead and imposed/live load at repair (column, 120 kN; beam, 30 kN).
4. Frame 4, control specimen with no repairs.

Strain and deflection monitoring devices were zeroed at the repair load. In each load cycle, the column load was increased or decreased in 10 kN increments followed by the beam load in 5 kN increments. Strains and deflections were recorded at each loading increment. A programme of loading and unloading cycles was conducted on each frame to simulate operation under design loads before finally loading to failure.

3 Stress analysis

In order to predict the stress distribution in the repaired section during various load configurations, plane frame analyses were conducted at sections coinciding with the experimental strain gauge positions, ie at the midspan of the beam and at the top of the repair in the column leg. Linear elastic behaviour was assumed throughout. At each load state subsequent to repair, the plane frame analysis was conducted with reduced member stiffnesses where appropriate. A solution was determined by trial and error in order to achieve a linear strain distribution across the section and with the required stress resultants. This involved varying the position of the neutral axis until the sum of the internal axial forces and moments was in equilibrium with the total forces and moments applied externally to the section. The assumptions used in this analysis are fully detailed in reference [2].

4 Material properties

The compressive modulus of the concrete and repair material were measured using 100 mm diameter cylinders. In the case of the repair material, these were cored from 100 mm thick test panels. The compressive modulus of the spray applied material averaged 16.6 kN/mm^2, which was lower than values of between 20 - 25 kN/mm^2 recorded by the manufacturer, and compare with 30.0 kN/mm^2 for the substrate concrete. This may be attributable to the awkward task of using equipment and materials designed for large volume repairs on construction sites for small repairs on model structures.

The adhesion of the repair material to the substrate concrete was assessed by applying the repair to concrete blocks 200 mm square which had been prepared either by water jetting or by grit blasting. One block was prepared by spraying downwards and another by spraying upwards on to the concrete substrate. After curing for 28 days, pull-off tests were conducted and the results are presented in Table 1. Little difference in bond strength was achieved with either surface preparation. However, the results would suggest that the bond of the sprayed material is not as good when sprayed upwards as when sprayed downwards. It should be noted that in many cases the limiting factor was often the tensile strength of the concrete. In general, excellent bond was achieved between the repair material and the substrate concrete.

Table 1. Pull-off strengths.

Surface treatment	Spray direction	Pull-off strength (N/mm$^{2)}$
water	down	1.80 (results
grit	down	1.84 (based on
water	up	1.02 (mean of
grit	up	1.14 (4 readings

5 Structural performance

Beam load against beam deflection curves provide a useful indicator of overall structural performance and these are shown in Figs 2, 3 and 4 for frames repaired under zero load, dead load only and dead plus imposed/live load, respectively. Examination of Fig. 2 shows the behaviour of the H-frame repaired under zero load to be similar to that of the control frame, although above 60 kN the slightly more flexible behaviour of the repaired frame may be reflecting its lower modulus of elasticity.

In Figs 3 and 4 there is a different performance up to beam loads of about 60 kN because some deflection has occurred before repair and this is not shown on the curves. However, above 60 kN, the load-deflection curves tend to merge and the effect of load level at the time of repair becomes insignificant. This is in accordance with the finding of Cairns [3].

The strain diagrams resulting from the stress analysis are overplotted with the experimental values for frame 1 in Fig. 5. This shows a linear experimental strain distribution across the section and provides strong evidence that full composite action is occurring across the repaired section for this repair material. Comparison of experimental and theoretical strain readings shows that, in practice, the neutral axis drops towards the centre of the section and the beam tends to behave as a homogenous section. This is due to the influence of the relatively high tensile strength repair material. Thus tensile strain predictions based upon the assumption of zero tensile strength are unduly conservative, at least at these lower levels of load.

In all cases, failures were characterised by yielding of the tension reinforcement preceeding crushing of the repair material (or concrete). The beam failure loads are summarised in Table 2. All beams exceeded the predicted ultimate capacity of 72.4 kN by a wide margin, possibly because there is still some capacity for moment transfer to the columns, despite the apparent formation of plastic hinges at the beam/column connections at beam loads of between 60 kN and 80 kN. Tensile crack patterns at failure were continuous through concrete and repair material with little evidence of bond failures at the interface, again supporting the view that effective composite action is taking place.

The lowest failure load of 95 kN for the frame repaired under full dead and imposed/live load may reflect the lower modulus of the repair material. This may result in a reduction of the proportion of the flexural compression load which it carries and place a higher stress on the concrete core of the beam. On the other hand, the lower modulus of this material may be beneficial in reducing the development of bond line stresses during the curing phase, thus avoiding possible delamination and failure at a later stage.

Pull-off tests were conducted on each frame at several repair locations after loading to failure. There appeared to be no significant difference in structural performance between the surface preparation techniques of water jetting and grit blasting.

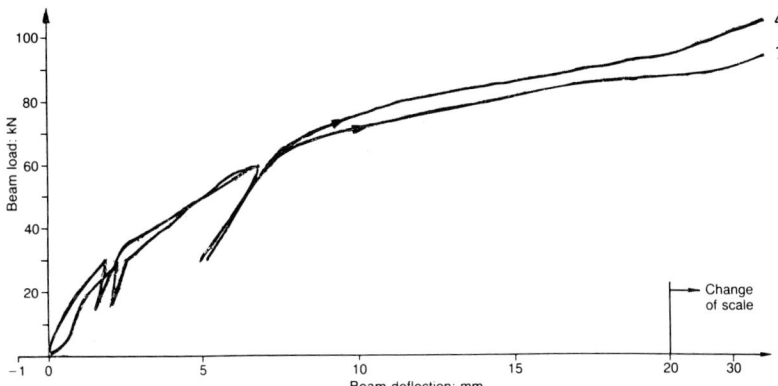

Fig. 2 Beam load against beam deflection, H-frames 1 and 4

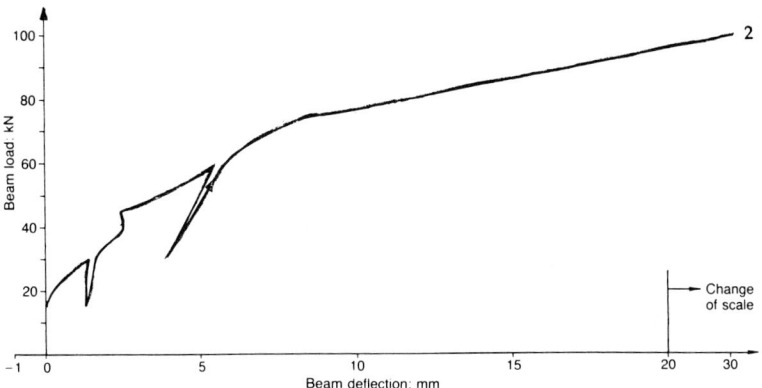

Fig. 3 Beam load against beam deflection, H-frame 2

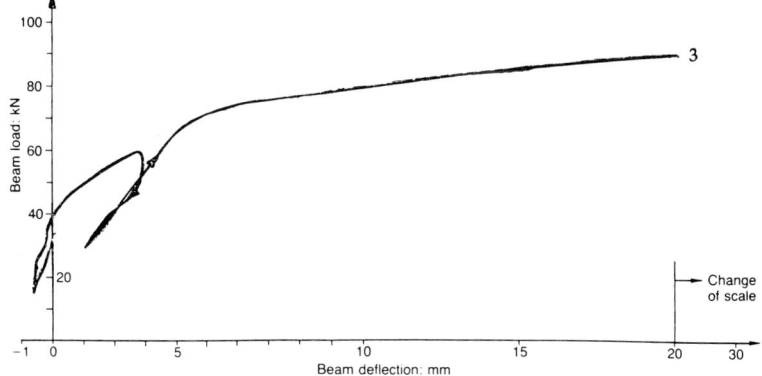

Fig. 4 Beam load against beam deflection, H-frame 3

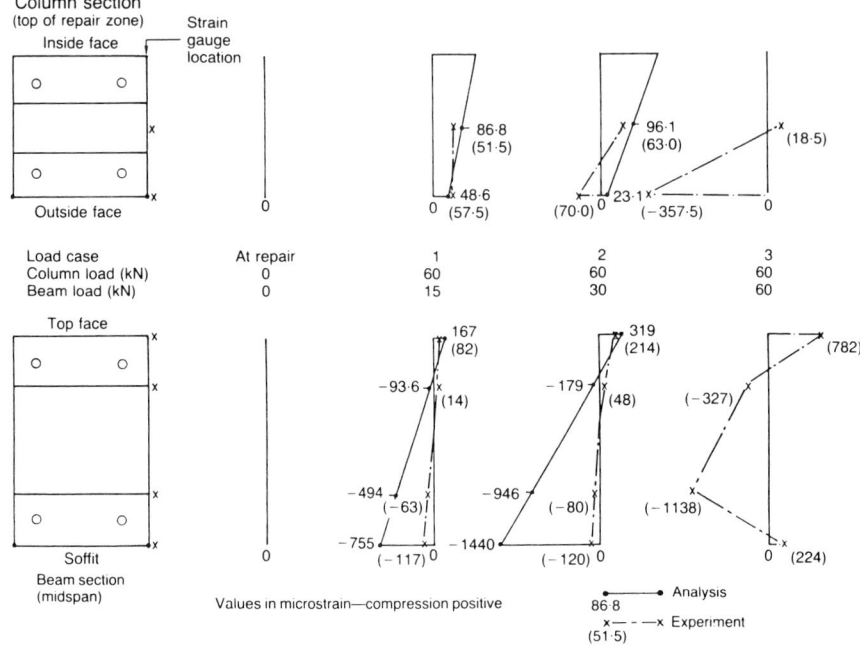

Fig. 5 Strain diagrams for frame 1.

Table 2. Summary of failure loads.

H-frame No.	Load at repair		Column load at beam failure (kN)	Beam failure	
	Column (kN)	Beam (kN)		Load	Mode*
1	0	0	60	100	PTF
2	60	15	60	105	PTF
3	120	30	60	95	PTF
4	N/A	N/A	60	110	PTF

6 Concluding remarks

1. Experimental studies have demonstrated that full composite action can be achieved across a section repaired with a spray applied polymer-modified cementitious material.
2. The conclusions of earlier studies have shown that it is desirable for the

3. modulus of elasticity of the repair material to lie within +/- 10 kN/mm² of the substrate concrete. The lower limit has been confirmed by the results of this work.
3. Water jetting and grit blasting of the substrate concrete both produced failures other than at the interface and are therefore suitable methods of surface preparation.
4. Load levels at the time of repair up to the full design load did not appear to influence the ultimate behaviour which, in all cases, was similar to that of the control unrepaired structure.
5. Stress analysis aimed at predicting the performance of repaired members under subsequent design loads may be based upon the assumption of linear elastic behaviour. However, the conventional procedure of neglecting the tensile strength of the concrete may be unduly conservative and produce an error in the position of the neutral axis.
6. The results of this study should provide designers with the confidence to use spray applied materials for large volume structural patch repair given an appropriate predictive analysis, appropriate materials and appropriate standards of workmanship.

7 Acknowledgements

The research described in this paper was undertaken with the support of the Department of the Environment, the Engineering and Physical Sciences Research Council, Balvac Whitley Moran Ltd., G. Maunsell & Partners, and Sika Ltd.

8 References

1. Mays, G.C. (1992) *Durability of Concrete Structures*, E & FN Spon, London.
2. Mays, G.C. and Barnes, R.A. (1995) The structural effectiveness of large volume patch repairs to concrete structures. *Proceedings Institution of Civil Engineers Structures & Buildings*, Vol.110, Nov., pp. 351-60.
3. Cairns, J. (1996) Alternative strategies for temporary support during structural repair of reinforced concrete beams. *ASCE Journal of Structural Engineering*, Vol. 122, No. 3, March, pp 238-46.
4. Emberson, N.K. & Mays, G.C. (1990) The significance of property mismatch in the patch repair of structural concrete. (1) Properties of repair systems. (2) Axially loaded reinforced concrete members. *Magazine of Concrete Research*, Vol. 42, No. 152, Sept., pp 147-70.
5. Emberson, N.K. & Mays, G.C. (to be published) The significance of property mismatch in the patch repair of structural concrete. (3) Reinforced concrete members in flexure. *Magazine of Concrete Research*.
6. British Standards Institution. (1985) *Structural Use of Concrete. Part 1: Code of practice for design and construction.* BSI, London. BS8110: Part 1.

22 LONG TERM PERFORMANCE OF SPRAYED CONCRETE REPAIR IN HIGHWAY STRUCTURES

P.S. Mangat and F.J. O'Flaherty
School of Construction, Sheffield Hallam University, Sheffield, UK

Abstract
The paper presents the preliminary findings of a project aimed at determining the long term performance of sprayed repair in highway structures. Different generic materials were used to repair deteriorating bridge structures using the dry spray process. The performance of these repairs was monitored to gain an understanding of the structural interaction between the spray applied repair material and the concrete substrate. The results show that the elastic modulus (of both the repair material and substrate) is the property which has the greatest effect on the performance of the repair.
Keywords: Bridge repair, material properties, performance, redistribution of stress, repair materials, sprayed concrete.

1 Introduction

Deterioration of reinforced concrete highway structures is a problem that is demanding increasing attention world-wide. This is due primarily to the corrosion of the reinforcing steel within the concrete, which becomes subjected to a severe corrosive environment containing chlorides and carbon dioxide. The most common form of rehabilitation for deteriorating structures is to replace the damaged concrete with a new material, thus increasing the design life of the structure.

Since its conception in 1895, sprayed concrete is extensively used world-wide as a repair technique for deteriorating concrete structures [1], but little information exists on its long term structural performance in a repair situation. Specification of sprayed repairs (and other repair techniques) at present in the UK is based on an inadequate understanding of the interaction between the substrate and the repair materials, especially in the long term when the effects of property mismatch becomes more pronounced. This oversight is addressed by Mangat and Limbachiya [2], where basic material properties (such as elastic modulus, shrinkage and creep) which influence long-term performance of concrete repair are investigated. Researchers elsewhere [3-6] have also investigated the significance of property mismatch between repair materials and concrete substrate. The current standard for specification of repairs, BD 27/86: Materials for the repair of Concrete Highway Structures [7], is based on the "best of current practice" with no consideration given to potential property mismatch between the repair material and the substrate.

Irrespective of which repair technique is employed, basic material properties such as modulus of elasticity, shrinkage and creep are arguably the most important properties which influence the long term performance of concrete repair. A mismatch of modulus of elasticity between the repair material and substrate can affect redistribution of stresses, whereas shrinkage and creep of the repair material may generate internal stresses. The effect of redistribution of these stresses on the performance of the repair needs to be investigated.

A project is currently underway which will determine the performance of concrete repair in highway structures. This involves repairing deteriorated concrete highway structures using different repair techniques and materials, and monitoring the performance of these repairs to determine their structural compatibility with the substrate concrete.

This paper presents the preliminary findings from repairs carried out using spray applied repair materials.

2 Details of bridge sites

Two deteriorating highway structures of different forms of construction have been used to monitor the performance of sprayed concrete repairs, namely, Gunthorpe Bridge carrying the A6097 in Nottinghamshire and Lawns Lane Bridge, near Wakefield, in West Yorkshire carrying part of the M1 just south of junction 42.

Gunthorpe Bridge is a three span, reinforced concrete arch bridge spanning the River Trent at Gunthorpe, Nottinghamshire. It was built in 1927 to replace an old iron toll-bridge due to an increase in heavy traffic using the bridge. The central arch in the bridge spans 38.1m, while the two side arches span 30.9m. Each of these three arches contain four ribs. An elevation of Gunthorpe Bridge is shown in Figure 1.

Lawns Lane Bridge is a three span reinforced concrete bridge which carries part of the M1 between junctions 41 and 42 in West Yorkshire. It was built in the mid 1960's and consists of insitu deck panels supported by prestressed beams, all of which are carried by reinforced concrete piers and abutments. A view of Lawns Lane Bridge is shown in Figure 2.

Fig. 1. View of Gunthorpe Bridge

3 Details of bridge repairs

3.1 Monitoring equipment

The redistribution of stress within the repair was monitored in order to determine the effectiveness of the repair patch in developing an efficient composite action with the substrate. This was achieved using vibrating wire strain gauges which were strategically placed on the concrete substrate, repair material and steel reinforcement. When the tension of the wire is increased in a vibrating wire strain gauge (i.e. as on a member with an increasing tensile stress), the frequency of the wire increases; alternatively it decreases in compression. Hence the change in strain can be calculated using the gauge equation:

$$\delta\varepsilon = k\,(f_1^2 - f_2^2) \tag{1}$$

Fig. 2. View of Lawns Lane Bridge

where $\delta\varepsilon$ is the strain change
k is the gauge constant (3×10^{-3})
f_1 is the datum frequency in Hertz
f_2 is the frequency of subsequent readings in Hertz

(Note: Positive sign indicates compressive strain; negative sign indicates tensile strain)

3.2 Location of repair areas

Three spray applied materials were instrumented at Gunthorpe Bridge. Figure 3(a) shows the positions of the repair areas on the south abutment at Gunthorpe Bridge. Each repair area was approximately 1.8m x 2.3m. Figures 3(b) and 3(c) shows the gauge configuration in the spray applied materials. Three gauges were positioned within the repair i.e. one on the cut back substrate (labelled "subs" in figures 3(b) and 3(c)), one welded to the steel reinforcement (labelled "steel") and one embedded within the body of the repair material (labelled "emb").

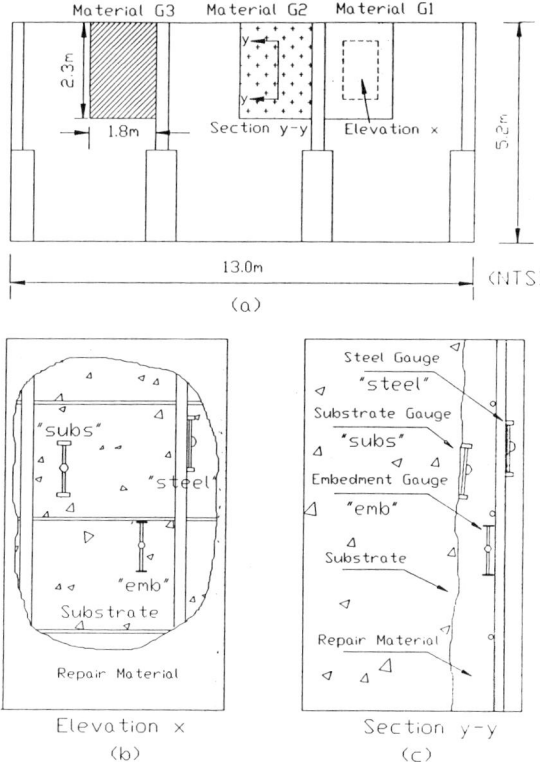

Fig. 3. Location of repair areas and position of gauges in the repair materials at Gunthorpe Bridge: (a) elevation of south abutment; (b) gauge configuration within a typical repair; (c) section through a typical repair.

Five spray applied materials were instrumented in total at Lawns Lane Bridge. Material L1 was applied on the north east pier, as shown in Figure 4(a), whereas material L2 was applied on the east face of the north west pier, also shown in Figure 4 (a). Three materials, L3, L4 and L5, as shown in Figure 4(b), were instrumented on the north abutment at Lawns Lane Bridge. The gauge configuration for monitoring the performance of the materials at Lawns Lane Bridge was similar to that employed on the materials on the south abutment at Gunthorpe Bridge, as shown in Figures 3(b) and 3(c), (i.e. gauges on the cut back substrate, steel reinforcement and embedded in the repair material). Where applicable, the adjacent substrate concrete was instrumented with a surface gauge to monitor any redistribution of stress which may occur.

4 Details of repair materials

Eight spray applied materials were monitored in total, three at Gunthorpe Bridge and five at Lawns Lane Bridge. The structural performance of three of these materials will be presented in this paper, material G1 from Gunthorpe Bridge and materials L1 and L4 from Lawns Lane Bridge. These three materials are commercially available and are supplied as single component systems requiring only the addition of clean water on site. The basic properties of the materials such as elastic modulus, shrinkage

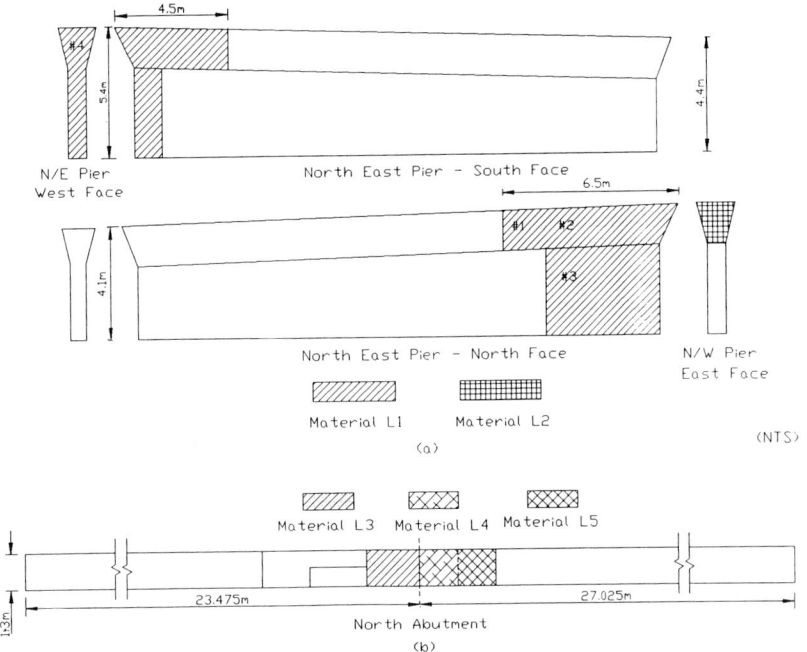

Fig. 4. Location of repair areas at Lawns Lane Bridge: (a) elevations of north east and west piers; (b) elevation of north abutment.

and creep were obtained from laboratory tests. Details of these spray applied repair materials are as follows:

Material G1 is a single component polymer modified cement based concrete mix. The material composition consists of rapid hardening Portland cement (minimum cement content of 400 kg/m^3) and 5 mm maximum sized graded limestone (clause 1702 of DoT Specification for Highway Works, Part 5). Silica-fume and a copolymer are also incorporated in the mix. This material achieves a 28 day compressive strength of 60 N/mm^2. It has a dry density of 2250 kg/m^3 and modulus of elasticity of 31.1 kN/mm^2.

Material L1 is a blend of low alkali Portland cement, silica-fume and high purity limestone aggregates together with a system of compatible admixtures. It is designed to meet the Department of Transport requirements for sprayed concrete as detailed in BD 27/86 [7] and clause 29F (Midlands Links Specification). The maximum aggregate size is 3 mm. When mixed at a 0.12 water:powder ratio, this material has a density of 2210 kg/m^3, modulus of elasticity of 22.7 kN/mm^2 and a 28 day compressive strength of 60 N/mm^2.

Material L4 is a factory blended dry spray gunite for the repair, strengthening or construction of concrete structures. This material composition consists of Portland cement, silica sand and admixtures including plastic fibres. The maximum aggregate size of the sand is 5 mm. This is designed to give a minimum 28 day compressive strength of 40 N/mm^2 when sprayed at a maximum water to cement ratio of 0.35. The modulus of elasticity, when tested in accordance with BS1881:Part 121:1983 [8] was 29.1 kN/mm^2.

5 Preliminary results and discussion

Figure 5 shows the measured strains, plotted at weekly intervals, in three of the spray applied materials under observation. Datum readings on each of the strain gauges were taken one day after the application of the repair material. Strain readings were linearly interpolated for weeks two and three as the automatic logging device was not installed until week four.

A detailed explanation of the measured strains in the repair materials as shown in Fig. 5 is outside the scope of this paper and will therefore be given in future publications. The aim of this paper is to emphasise that the elastic modulus (of both the repair material and substrate) is the most important property to be considered when selecting a repair material for concrete repairs.

Fig. 5 (a) shows the data obtained from the gauges in material L4 at Lawns Lane Bridge. These strains are primarily caused by deformation (shrinkage and creep) of the repair material. The deformation in the substrate due to external loading had already occurred prior to instrumentation. Material L4 has a stiffness which is greater than the stiffness of the substrate (E_{repair} = 29.1 kN/mm^2, E_{subs} = 23.8 kN/mm^2). As a result, the repair material exhibits net shortening (compressive strain), thereby transferring compression into the substrate and steel reinforcement (see Fig. 5 (a)). This is typical for all materials at Lawns Lane Bridge where the stiffness of the repair material is greater than the stiffness of the substrate. It has also been observed that in

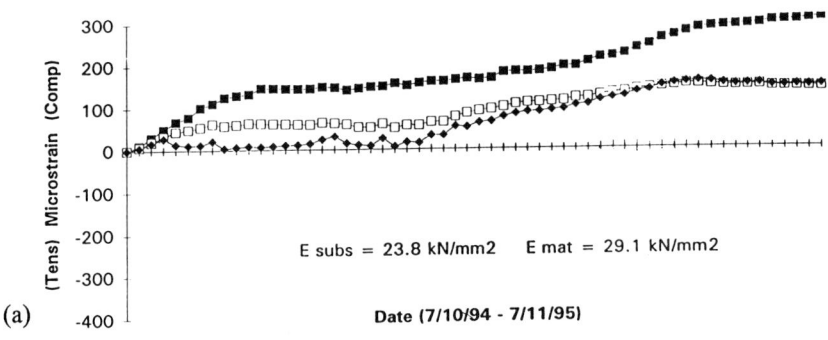

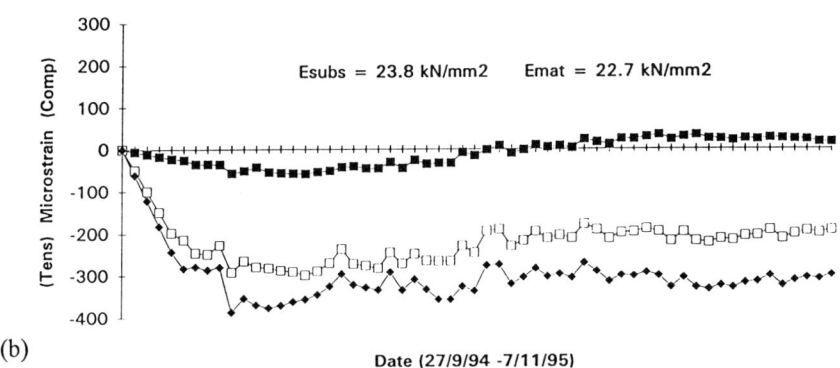

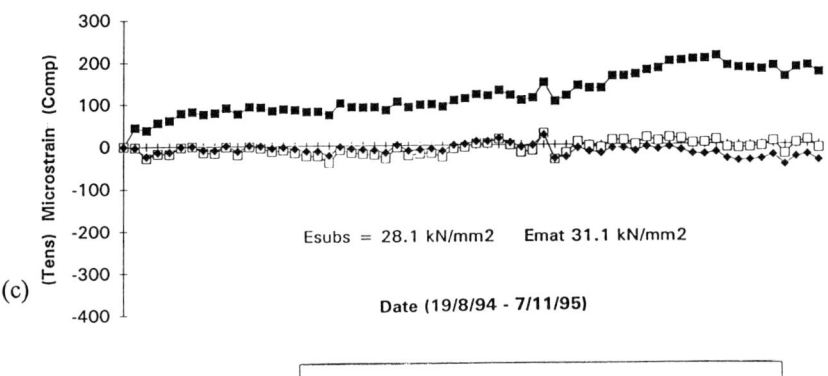

fig. 5. Redistribution of stress in three different sprayed materials:
(a) Material L4; (b) Material L1; (c) Material G1

the long-term, repair materials which are stiffer than the substrate allow greater redistribution of externally applied stress to take place from the substrate to the repair patch.

A different situation occurs for repair material L1 whose stiffness is less than the stiffness of the substrate (E_{repair} = 22.7 kN/mm², E_{subs} = 23.8 kN/mm²). Since the substrate is marginally stiffer than the repair material, it remains unaffected by the shrinkage of the repair. Consequently, there is little strain in it. In fact, it restrains the shrinkage of the repair effectively and this together with any initial expansion of the repair material at very early age causes net tension in the repair patch. This effect on the repair patch is also transferred to the steel reinforcement, which also exhibits some tension - significantly less than the repair material owing to the higher stiffness of the steel. These tensile strains were responsible for the catastrophic cracking which occurred in material L1 at Lawns Lane Bridge, see Fig. 6). It has also been observed that a material in this situation is unable to attract external applied stress from the substrate.

Fig. 5 (c) shows the strains measured in material G1 at Gunthorpe Bridge. The stiffness of the material is 31.1 kN/mm², whereas the stiffness of the substrate is 28.1 kN/mm². Compressive stress is transferred to the substrate due net shortening (compressive strain) in the repair material (similar to material L4 at Lawns Lane Bridge, see Fig. 5 (a)), but in this instance, the effect on the steel reinforcement is negligible. This may be due to the fact that the difference in stiffness between material G1 and the substrate is only 3 kN/mm² at Gunthorpe Bridge, whereas the difference between material L4 and the substrate is 5.3 kN/mm² at Lawns Lane Bridge. This would lead to a lower transfer of stress from the repair material to the steel reinforcement.

Fig. 6 View of cracking in material L1 at Lawns Lane Bridge

6 Interim conclusions

The following conclusions are based on the preliminary findings obtained from the different spray applied materials under test:

- The elastic modulus is the most important property to be considered when selecting a repair material for concrete repairs
- In a situation where the elastic modulus of the repair material is greater than the elastic modulus of the substrate, redistributed stresses within the repair are predominately compressive.
- A material which has an elastic modulus which is greater than the elastic modulus of the substrate will attract load from the parent concrete.
- A material which has an elastic modulus which is less than the elastic modulus of the substrate will not attract load from the parent concrete.
- Severe cracking may occur in a repair material when the stiffness of the material is less than the stiffness of the substrate.

7 Acknowledgements

This paper presents some of the results of a research project funded by the LINK TIO Programme entitled "Long term performance of concrete repair in highway structures". The partners in this project are: Sheffield Hallam University, V.A. Crookes (Contracts) Ltd., Flexcrete Ltd., M.J. Gleeson Group plc. and Department of Transport. The authors also acknowledge the support of Nottinghamshire County Council and The Highways Agency through their co-operation with the field studies.

8 References

1. Shaw, M. (1992) Structural concrete repair: Machine applied systems. *Construction and Building Materials*, Vol. 6, No.1, pp. 47-51.
2. Mangat, P.S. and Limbachiya, M.K. (1995) Repair material properties which influence long-term performance of concrete structures. *Construction and Building Materials*, Vol. 9, No. 2. pp.81-90.
3. Hewlett, P.C. and Hurley, S.A. (1985) The consequences of polymer-cement mismatch. *Design Life of Buildings*, Thomas Telford, London, pp.179-196.
4. Emberson, N.K. and Mays, G.C. (1987) Polymer mortars for the repair of structural concrete: the significance of property mismatch. *Proc. ICPIC '87: 5th Int. Cong. on Polymers in Concrete. Brighton, September 1987, pp. 235-242.*
5. Plum, D.R. (1989) Repair materials and repaired structures in varying environment. *Proc. 3rd Int. Seminar on the life of Structures - The Role of Physical Testing. Brighton.*

6. Wood, J.G.M. *et al.* (1989) Defining the properties of concrete repair materials for effective structural application. *Proc. Structural Faults Repairs '89,* Vol. 2, London, pp 231-236.
7. Department of Transport. (1986) *Materials for the Repair of Concrete Highway Structures.* Department of Transport Highways and Traffic Departmental Standard. London. BD 27/86.
8. British Standards Institution. (1983) *Testing Concrete. Part 121: Method for determination of static modulus of elasticity in compression.* BSI, London. BS1881: Part 1.

PART SIX
SPECIFICATION

23 CONTRACTUAL ASPECTS OF TESTING SHOTCRETE AND ROCKBOLTS

R.B. Clay and A.P. Takacs
Dar Al Handasah, Adana, Turkey

Abstract
Shotcrete is frequently used, with rock bolts as support for cut slopes and in tunnels. For shotcrete, tests generally specified include laboratory tests, suitability tests, and quality control tests, as well as proficiency testing of nozzlemen. For rockbolts, two types of testing are generally specified -- proof tests, required before bolts may be used; and production tests, performed on bolts installed in the permanent works.

This paper summarises some practical problems which may be encountered in applying and interpreting the documents, giving examples from one particular contract. It discusses the suitability and relevance of some often-prescribed tests, and offers suggestions for drafting future specifications.
Keywords: shotcrete, rockbolts, testing, NATM, tunnel, sprayed concrete, specification, contract.

1 Introduction

The majority of shotcrete is placed to support tunnels and cut slopes of hillsides, and it is generally used in conjunction with rockbolts. It is therefore appropriate to give some thought to these as well.

The authors each have experience of many tunnelling projects in various parts of the world. They are currently based in Turkey, supervising the driving of nearly five route km of twin three-lane motorway tunnels, with eight portals, each with a large cut face. Work is still in progress, and involves these estimated total theoretical quantities:
Tunnels: 300, 000 m² of shotcrete and 300,000 rockbolts; Portals: 21,600 m² of shotcrete and 30,000 rockbolts. This paper is based upon their experience, and the

current project is used only as an example. The views expressed herein are solely those of the authors. One such is 'When you are up to your neck in alligators, still try to remember that you came to drain the swamp.' It is always relevant in any construction job to bear in mind the main purpose. Here, it is to enable traffic to flow. Testing should be done towards that aim.

The current project includes several viaducts, some of steel, most of pre-cast pre-stressed concrete. Steel production, whether for fabrication or rock-bolts, is a highly technical and tightly controlled process. Concrete for the pre-stressed beams is produced in sophisticated batching plants, with ingredients accurately dispensed, and concrete is placed under closely controlled conditions. The production of shotcrete, however, is less scientific, as will be further discussed below.

The viaducts are designed using well-known principles, applying a system of loading in a straight-forward application of well-proven formulae. Assumptions are made of the loading to be imposed on the ground; and tests will be carried out to establish its load-bearing capacity. The design of shotcrete and of rockbolts, however, is quite different.

In tunnelling, and in the creation of cut slopes, the concern is with what loads the ground will exert when the natural support is removed. This problem does not lend itself to easy prior resolution. Hence tunnelling is considered an art, not a science; and especially so when using a support system based on the use of shotcrete and rockbolts. The main element of support is the ground itself; the shotcrete and the rockbolts are only required in order to enable the ground to be self-supporting. Consequently, it is not possible to forecast accurately the actual loading on the support.

The load-bearing capacity of the support as a whole is not calculable, due to the varying thickness of the applied shotcrete, due to the strength increase in relation to the increasing load applied by the ground, and due to the gain in strength of the rockbolt grout, which allows transfer of load from ground to rockbolt. Too rigid a support at too early a stage may be no better than too flexible a support, since the support must then carry the entire loading, without mobilising the load-bearing capacity of the ground.

The engineer must decide from time to time what support to apply. Only as the load comes on does it become possible for him to judge whether the support is adequate, and likely to remain so. By then, it is too late to install different primary support, and emergency measures have to be taken; a process rather like judging a recipe by the amount left on the diners' plates. The success of the design, on the other hand, is judged solely by visual inspection and by measuring deformation. Nothing will reveal that the support is too strong, and there is therefore little incentive on most contracts even to try to reduce the support.

2 Shotcrete in general

Shotcrete provides support for the rock, to replace partly the support which has been removed, and to create a smooth continuous load path. It also holds chunks of rock in place, thus enabling them still to serve a structural purpose. A minimum thickness of shell is thus necessary, and this acts in both shear and in compression. The thickness of a layer of shotcrete required to hold loose rock in place is actually quite small - a few millimetres. When shotcrete was first introduced for this purpose, a 25 mm layer thickness was quite usual, but layers are now about five times thicker.

In every contract, the quality and testing of the materials will be closely specified. In the kitchen, it is difficult to produce an unacceptable dish if the ingredients include cream, chocolate and rum. But with shotcrete, starting with good materials may or may not result in a good end product; but a good end product cannot be achieved with poor materials. The quality of the basic ingredients must comply with the specification. Our concern here is with other factors, the most significant of which is the nozzleman. Upon him rests the responsibility for the application and for the water/cement ratio, the most significant factor affecting the strength of the shotcrete. Here, we consider principally the dry shotcrete process, whereby the water is added to the mix as it leaves the nozzle.

3. Shotcrete quality and quantity

The following are fairly typical requirements of a specification:

1. Mix design -- water/cement ratio: The water content has to be controlled by the nozzleman to suit the conditions of the shotcreting surface and location of application. An indication that the water/cement ratio is in the correct range is that the shotcrete seems to have a slightly shiny appearance immediately following application.
2. Placing of shotcrete: Measures to establish the total thickness of shotcrete are usually required. These may include visual guides installed before shotcreting or holes drilled after completion of shotcreting.
3. Quality control tests -- in-situ compressive strength: A usual method of testing is for test panels to be sprayed at the same time as production shotcrete is being placed, perhaps one panel for every 100 cubic metres of shotcrete applied. Several cores are then taken from the panel, for testing at various ages. As a control, it is common also occasionally to drill test cores from the in-situ shotcrete.
4. Qualification of operators: Every operator should be trained, and tested by the spraying of test panels, to be tested and approved before he is employed to apply production shotcrete. Such qualification is necessary not only for nozzlemen, but also for the plant operators.
5. Placing of shotcrete: The optimum distance between nozzle and surface of application is normally taken as 1.0 to 1.3 metres. The nozzle should be positioned at right angles to the surface of application. Both of these measures serve to ensure good compaction, and to reduce re-bound.

Rock breaks to a ragged profile, raggedness which may be reduced by careful blasting, or, in the extreme, by using a TBM. The final shotcreted surface, however, is required to be relatively smooth. Shotcrete thickness therefore varies considerably, from the barest minimum at high spots to quite substantial thicknesses where the rock has broken back to natural joints. It is normal for a minimum thickness to be specified. However, it is not possible to measure this minimum thickness after the shotcrete has been applied, unless pins are used, fixed to the high spots before shotcreting starts, but this is a rare practice. When shotcrete is sprayed into place, some bounces off; the re-use of such rebound is generally prohibited. The proportion of rebound depends principally on the skill of the nozzleman, but may be 50% or more. It may sometimes be possible, although rarely practicable, to measure the quantity of this rebound, and thus deduce the mean thickness of applied shotcrete. In practice this is never done, and the shotcrete thickness is taken as a mean of measurements taken through drill holes. Even these, though, are not very accurate, and their location is virtually at random for establishing the true thickness of shotcrete, a figure only really of significance for payment. From a technical viewpoint, the amount of shotcrete to be applied remains a matter of judgement -- does it appear sufficient? To provide proper support, it is necessary that enough shotcrete is placed to smooth out the discontinuities of the broken rock face, and to provide shear resistance to pieces of rock which otherwise might fall and possibly lead to unravelling of the rock formation.

4. Shotcrete failure, its significance and the avoidance of failure

The only tests on the completed shotcrete in the tunnel are compressive tests and measurement of the thickness. The shotcrete may fail the strength test, or prove to be not as thick as specified. Despite such failure, the shotcrete may be perfectly adequate for its intended purpose. It is quite normal for the specification to require that, if the shotcrete fails either test, extra shotcrete be placed in the tunnel to increase the strength of the support. However, if the support is adequate, then this extra support is unnecessary. In addition, such application of extra support may subsequently have to be trimmed away to obtain the required clearances inside the tunnel. If this is likely to be the case, then it may be a requirement of the specification that the existing support be removed and replaced, even though it is structurally adequate but happens not to comply with the specification. It is in situations such as these that it is well to keep a sense of proportion, and to remember the primary purpose of the contract.

As well as the above tests, deformation measurements might show that the applied loads are more than the shotcrete can take, or this may be deduced from observing cracking of the shotcrete. In such cases the shotcrete is not sufficient to resist the loading, even though it may have passed the strength tests, and be of adequate thickness. It may prove necessary to remove such support, trim the ground, and apply a stronger system of support, despite the shotcrete complying with the specification with regard to both strength and thickness.

Only good materials must be used for making shotcrete, but good workmanship is even more important. Although the specification may include guidance on method, this is not often followed in practice unless the nozzle is remotely operated; but then the water content, which relies upon visual inspection, may be even less precise. The job of nozzleman is not a pleasant one sought by the best workmen. Even with the most enlightened of employers, proper protection, ventilation and lighting for the nozzleman is extremely rare. It is interesting to contrast the working conditions and remuneration, and the consequences and costs of failure of a nozzleman and of a chef. If the top man on site were to be required to be certificated as a nozzleman, then conditions might be improved to such an extent that consistently good shotcrete would always be properly applied. To be realistic, shotcrete as applied generally turns out to be more than adequate for its purpose. The contractor or designer may well consider it cheaper to make provision for extra excavation and extra thickness of shotcrete than to improve the working conditions.

5 Rockbolts

5.1 Types of rockbolt
Standard lengths:

- For tunnels: 4m (80%), 6m (17%), 8m (3%).
- For slopes: 4m (5%), 6m (15%), 8m (20%), 12m (40%), 16m (10%), 18m (5%), 22m and 24m (5%).

Standard types:

- SN (Stor Norfors), PG (post grouted), IBO (injection bore bolts)
- All bolts are fully grouted and untensioned.
- In tunnels, 90% are SN bolts, 10% are IBO, PG not used.
  On slopes, 50% are PG, 30% are SN and 20% are IBO.

5.2 Rockbolt proof tests
Earlier in the contract, proof tests were performed in various ways and using differing procedures. In 1994, the method proposed by the ISRM [4] was included in the new technical specification. A principal requirement of the ISRM method is that the rockbolts must be loaded progressively and the corresponding deformations measured until failure occur. Tests are required to be carried out on 'each type' and for each 'different geological condition'. In addition the technical specification requires the testing of at least five rockbolts of each type. If we combine these requirements with the assumed geological conditions (Rock Classes 3, 4, 5 and 6) the tests required are:

Table 1. Rockbolt tests required

	RC 3	Rock Class 4		Rock Class 5			Rock Class 6		
Length	4 m	4 m	6 m	4 m	6 m	8 m	4 m	6 m	8 m
Type									
SN	*	*	*	*	*	*	*	*	*
IBO	-	*	*	*	*	*	*	*	*

Total 19 tests x 5 bolts each = 95 tests to be performed. However, there is still some ambiguity regarding:

- 'different types' -- does a change of length represent a different type?
- 'different geological conditions' -- it is not possible to find all rock classes in a defined surface.
- may the test be performed on the main slope surface? Or must it be nearby?
- is it necessary to test types such as PG, which are less often used in practice?

During 1994, 67 rock bolts were tested, ranging from 0.50m to 24m. Although the minimum length of rockbolt to be used was 3m, it was found that, regardless of length, failure always occurred in the steel bar, at its ultimate tensile strength. These tests thus revealed little about the anchorage. It was therefore decided to reduce the length of the test bolts to ensure that the failure occurred between grout and steel and/or between grout and ground. The results of those tests have been presented in reference [3].

5.3 Proof tests at the east portal of the Ayran tunnel

One of the latest series of tests to be performed was required for the east portal slope of the Ayran tunnel, a permanent cut slope about 35 metres high. Alternative designs were considered for support of the rock debris, or tunnelling either partly in debris or all in sound rock. The second solution was chosen, requiring a heavier support for the slope itself, but increasing the safety of the tunnel drive.

An agreement was made between the contractor and the designer on the one hand and the engineer on the other, about what tests should be performed. Only SN type bolts were to be tested, made from 28mm and 32mm diameter rebar, and three bolts were to be tested for each length.

Table 2. Numbers of bolts to be tested (45 in all)

Geology	Diameter [mm.]	Length (metres)						Total
		0.25	0.50	0.75	1.00	1.25	1.50	
Rock Debris	⌀28	-	3	3	3	3	3	
	⌀32	-	3	3	3	3	-	27
Rock	⌀28	3	3	3	-	-	-	
	⌀32	3	3	3	-	-	-	18

Part of the permanent slope section was cut to the designed inclination and supported by shotcrete and mesh in accordance with the design. On this surface the 1.00, 1.25 and 1.50m long rockbolts were installed at the designed angle. On a different horizontal surface the 0.50 and 0.75m long rockbolts were installed vertically, to ensure the required grouted length. Because of the deep location of bed-rock, rockbolts for the rock were installed inclined on a nearby rock surface, prepared as above.

The bedrock, formed by alternations of sandstone and siltstone, is covered by slope wash and scree deposit of considerable thickness. This rock debris comprises mainly sandstone and quartzite in a matrix of light brown silty sand. Sound rock consisted of interbedded sandstone and siltstone conglomerate.

The theoretical bond length was calculated for each:

Table 3. Calculated theoretical rockbolt bond lengths

Rockbolt \ Geology	Rock debris		Rock	
diameter [mm.]	⌀28	⌀32	⌀28	⌀32
bond length [m.]	0.6-0.8	0.8-1.2	0.20	0.26

The surface preparation, the drill holes and installation of bolts were performed by a newly established subcontractor with only limited experience in the use of shotcrete and rockbolts. The strength of the grout was 23.3, 30.8 and 37.3 N/mm for rock debris and 21.8 and 26.0 N/mm for the rock. Specified minimum strength for grout is 20 N/mm. All holes were drilled 68 mm in diameter. Of the 45 tests which were to be performed, five are not included in the results. Two of the bolts were not tested, one

SNØ28-1.50m in debris and one SNØ32-0.50m in rock. From the test results of two other bolts, SNØ32-0.25m, and SNØ32-0.75m, it is evident that they were 'locked', and only the rebar was tensioned. A fifth bolt, SNØ32-0.50m was not properly grouted and failed at less than 20 kN pull force. There are thus 25 test results for rock debris and 15 for rock available for further analysis. For each bolt a load-deformation curve was plotted, and a test report prepared.

The results were re-analysed, partly confirming the previous conclusions. The load-deformation curves were re-plotted to give more accurate diagrams and the load-carrying capacity was thus more easily defined. Two aspects deserve emphasis:

- The number of rockbolts tested is insufficient to be regarded as conclusive.
- Not all test results can be regarded as accurate.

Considering previous test results, and on the assumption that longer bond length has higher carrying capacity, the results diagrams were adjusted. The calculated bond length is compared with the site results, as shown in Figure 1.

The 0.25m long rockbolts show a carrying capacity less than expected. One reason for this may be the different grouted lengths (which practically are difficult to achieve with high accuracy and strict definition). Another reason may be that the grout nearest the mouth of the hole is not transferring the load to the rock, the bond here being destroyed by the high tensile force transferred by the bolt itself. Accordingly we may consider that the actual bond length was less than 0.25 m. Results obtained in the rock debris are generally less than the calculated values. This may be explained by the high tendency for the walls of the drillholes to collapse, by the difficulty of cleaning the hole and by the inexperience of the subcontractor. During the execution phase, many PG type rock bolts were replaced by IBO type bolts, which install better in these conditions.

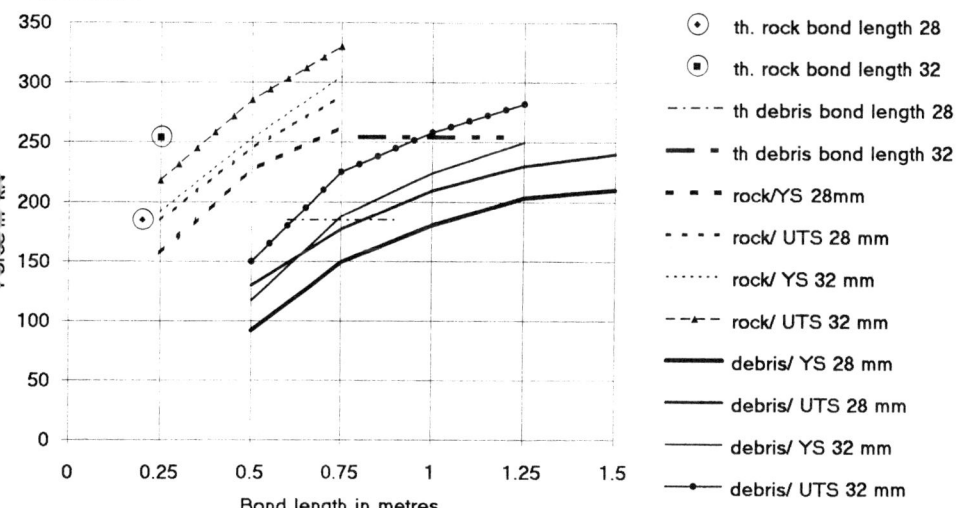

Fig. 1. Relation between bond length and carrying capacity ("th"=theoretical)

6 Conclusions

6.1 Shotcrete
Practical problems occur on site in applying the specification, which lays down methods of working. These are frequently unrealistic, and can neither be complied with, nor insisted upon. The writer of the specification needs to have an appreciation of the practical difficulties, and consideration for the workmen involved.

The suitability and relevance of the prescribed tests are open to question. Whilst it remains important to ensure that the materials used are of the required quality, the actual strength of the shotcrete and its actual thickness are of less importance than the answer to the question 'Are they adequate for the purpose?'. No test other than observation has yet been devised to establish this.

6.2 Rock bolts
The number of rock bolts to be proof-tested per type is not sufficient for statistical interpretation. To allow a meaningful conclusion to be drawn, the number of tests needs to be of the order of hundreds, which is simply not practical nor economic in the case of small projects. The tests take a significant time, which is likely to be under-estimated by the contractor. The testing procedure is not complicated, but requires a qualified experienced and professional team, unlikely to be available at the start of a new job, when shotcrete and rockbolts are most likely to be required. The proof test is actually not of very great significance. Results of such tests might influence the design, but generally this is complete well before test results become available. On such a small scale as usually obtains, the test results may only confirm the predictions, but will not be sufficient to contradict them. It is most unlikely that the results of the tests will be used to justify any reduction in the number or length of the rock bolts, even if this could have a significant influence on the cost of the project.

The tests required during construction have only been mentioned in passing. They absorb much time and money; but are they worth it? As a practical result of the proof test, it may be concluded that failures will occur only in rockbolts that are not properly grouted.

All that a successful test can indicate is that there is sufficient bonded anchorage length somewhere along the bolt. This may mean, for example, that a twenty-four metre long bolt passes the test, but is completely ungrouted except for a short length at the mouth of the hole. Such a bolt would serve no useful tensile function.

7 References

1. Austrian Concrete Society (1990) *Guideline on Shotcrete,* ACS, Vienna
2. Turkish Standard *TS 1246*
3. Takacs, A.P. (1994) *Testul de verificare prin întindere a ancorelor la tunelul Kîzlaç A din Turcia* Constructii Subterane, (ed. Romanian Tunnelling Association), Issue No. 5-6, pp. 5-7.
4. Brown, E.T. (Editor) (1981) *Rock Characterization, Testing and Monitoring -- ISRM suggested Methods: Document No. 2, Suggested methods for rockbolt testing,* International Society for Rock Mechanics, Pergamon Press, Oxford.

24 SHOTCRETE STANDARDS – AN AMERICAN PERSPECTIVE

I.L. Glassgold
Masonry Resurfacing & Construction Co Inc, Baltimore, USA

Abstract
This paper describes specifications and testing methods currently standardized in United States shotcrete practice. A brief review of standards development, in general, and shotcrete technology specifically, is presented. The division of responsibility between the two standards writing organizations, ASTM and ACI, is outlined. Current practice includes specifications and tests for both wet and dry process shotcrete. Some standards pertain directly to shotcrete, however, with some modification, many of the available ASTM concrete standards can be adapted to determine various shotcrete properties. Finally, future needs for shotcrete standards, especially in the plastic state are covered.

Keywords: Shotcrete, standards, wet-process, dry-process, specification, testing

1 **Defining Standards**

Our understanding and application of the concept of standards is loosely based on Webster's dictionary definiion of the generic term 'standard' - "something that is established by authority, custom or general consent as a model or example to be followed." This definition encompasses both social and technological standards and applies to their creation as well as their implementation. In the United States, there are some 400 organizations that write technical standards, based on this definition. When adopted by the appropriate authority, a standard may have the force of law and all that it implies.

2 **Background**

Social standards have influenced the growth of civilization from man's earliest beginnings. One of the earliest documented set of standards is the Code of Hammurabi which dates back some 5000 years and was a codification of Babylonian law, defining man's social conduct. The Decalogue of the Bible, couched in imperative terms is another set of similar social standards dating to the time of Moses, some 3200 years ago.

Modern archaeological techniques have uncovered ancient artifacts whose uniformity of material and construction indicate that technical standards must have existed in those days. Clay pottery from different societies is a prime example; as was the unique Roman brick that found its way into all corners of the Roman Empire. Nevertheless, until the recent past the nature of society and commerce was such that the need for extensive formal technical standards was very limited.

The advent of the Industrial Revolution in the late 18th and 19th centuries changed a mainly agrarian society to one of manufacturing and mass production requiring new means of transportation and communication and more efficient energy sources. The buzz words of this new age were "science", "invention", and "industry", as were the terms "free trade", "laissez-faire" and "caveat emptor." Mass production and mass marketing with all their virtues and evils soon led to abuses in the market place. It soon became apparent that there was a deep need for creating technical standards in a burgeoning technological society. They were needed to promote fair and equitable competition, product safety, manufacturing efficiency, commodity quality and technological advancement.

3 Concrete/Shotcrete BeginningsS

Coincidental with the rapid change to an industrial society came the development and growth of the concrete industry. The invention of Portland cement by Joseph Aspdin in England in 1824 became the catalyst that sparked modern concrete technology. By the end of the 19th century, steel reinforced concrete had been developed both in the United States and Europe. After importing cement from England for many years, cement became readily available in the United States when cement manufacturing began in Pennsylvania in 1871. By the early 1900's portland cement concrete had become an important factor in both heavy and light types of building construction.

4 Gunite/Shotcrete Technologies

The rapid acceptance of this new concrete technology led to innovations in concrete mixing, transportation and placement. An example of a new placement technique was the invention of the "cement gun" and its cement product called "Gunite" by Carl Akeley in 1911. It led to the development of the shotcrete industry as we know it today. The Gunite process utilized a device which introduced a dry cement-sand mixture into a moving stream of compressed air passing through a placement hose. Water was added to the mixture just prior to exit from the hose to produce a concrete mortar. From its introduction to the present day, this dry-mix process has remained essentially the same except for the development of continuous feed guns and the utilization of admixtures.

The wet-mix process was first mentioned in the concrete literature in 1916 several years after the introduction of "Gunite" but received little publicity or use thereafter. At that time the wet process premixed all its ingredients including water and utilized a pressurized tank to move the mortar or concrete through a placement hose with compressed air being added at the nozzle. The failure of the wet process to gain a foothold in construction was probably due to the fact that the equipment was not standardized or promoted adequately as was the Gunite process. We therefore hear very little of the wet process until the introduction in the 1950's of the True Gun, a dual-tank pneumatic device which provided continuous flow of the wet mixture. Thereafter, with the adoption of various types of concrete pumps, the wet process finally became a viable and economical application process.

In order to provide a generic substitute for the term Gunite and other trade names such as Blocrete, Guncrete, and Jetcrete, etal, around 1930, the American Railway Engineering Association (AREA) introduced the term "shotcrete" in the 1930's to describe the dry-mix process. In the United States, the term shotcrete refers to both dry and wet processes and is defined by The American Concrete Institute (ACI) as "Mortar or concrete pneumatically projected at high velocity onto a surface."

5 Standards Sources

From the very beginning gunite/shotcrete has been considered part of portland cement technology with the majority of the articles on the subject being published in publications devoted to concrete technology. In the United States there are two standards writing organizations which produce the majority of the standards for concrete construction 1) The American Concrete Institute (ACI), now known as ACI International and 2) The American Society for Testing and Materials currently known as ASTM. Between them they cover the entire field of concrete construction. Each had its origin at the turn of the past century because of the urgent need to standardize materials, manufactured products and processes.

5.1 ACI

ACI had its start in the early days of concrete technology because "a competitive market with a serious lack of standard practice in making concrete block had resulted in conditions so unsatisfactory that by the summer of 1904" a group of interested parties undertook the formation of an organization to discuss and correct the problem. It was called the National Association of Cement Users (NACU). The scope of the organization was soon extended to include all the various uses of cement. In 1913, its name was changed to the American Concrete Institute to more fully describe the aims and interests of the group.

From the very beginning, early articles on the subject of Gunite were published by NACU. ACI has continued this effort and is probably the foremost publisher of shotcrete documents in the world today.

In addition, in 1942, ACI formed its committee 805 on Pnuematically Placed Mortar that authored and published the standard, "Recommended Practice for the Application

of Mortar by Pneumatic Pressure" (ACI 805-51). By 1957 there was need to revise the aforementioned standard to reflect the many changes that had occurred in shotcrete technology. ACI Committee 506 (ACI 506) was organized to accomplish this task. Since its start in 1957, ACI 506 has sponsored many symposia and seminars in addition to many reports and standards some of which are listed below.
. Guide to Shotcrete, (ACI 506-85) Rev. 1990
. Shotcreting, SP-14 a shotcrete compilation
. Specification for Materials, Proportioning and Application of Shotcrete (ACI 506.2-77)
. Guide to Certification of Shotcrete Nozzlemen (ACI 506.3R-91)
. State of the Art on Fiber Reinforced Shotcrete (ACI 506.1R-84)
. Application and Use of Shotcrete, ACI Compilation #6 (1981)

All ACI documents, standards and non-standards are produced on a <u>democratic</u> <u>concensus</u> basis with input from volunteers representing contractors, manufacturers, academia, users, consultants, and the government. Also, geographic balance is an important factor in committee membership.

5.2 ASTM

In a similar manner ASTM began its work in the early days of the developing concrete industry. In 1898, the American section of the International Association for Testing of Materials was formed which was the forerunner of the American Society for Testing and Materials. From the very beginning ASTM had an abiding interest in steel and cement, two important ingredients in both concrete and shotcrete construction. However, until 1988, ASTM's involvement in shotcrete standards was very limited. Since 1988, ASTM Committee C09 on concrete has produced five shotcrete standards as follows:
. ASTM Test Method for Time of Setting of Portland Cement Pastes Containing Accelerating Admixtures for Shotcrete by Use of Gilmore Needles (C1102-1988).
. ASTM Test Method for Time of Setting of Shotcrete Mixtures by Penetration Resistance (C1117-1989)
. ASTM Specification of Fiber Reinforced Concrete and Shotcrete (C1116-1989)
. ASTM Specification for Admixtures for Shotcrete (C1141-1990)
. ASTM Practice for Preparing and Testing Specimens from Shotcrete Test Panels (C1140-1990)

Aside from these five, most ASTM standards used in shotcrete work are concrete oriented.

In 1990, ASTM established subcommittee C09.46 on Shotcrete to review and codify ASTM standards currently being used for shotcrete and also create new standards where necessary. In the United States responsibility for creating and maintaining standards for concrete and shotcrete are shared by ACI and ASTM. Other technical societies with specific interests in Shotcrete have established their own standards. For instance, AREA has established a shotcrete standard for the railroad industry while the American Water Works Association (AWWA) has their shotcrete specification for use on water pipe lining and coatings. In addition, within ASTM, Committee C08 on Refractories has established standards based on cold nozzle mix gunning which is their term for shotcrete.

In 1936, in order to avoid conflict and duplication, ACI and ASTM developed a memorandum of understanding in connection with reinforced concrete standards. ASTM would address Material specifications and test methods while ACI would handle matters pertaining to concrete design and construction practice.

6 Properties of Shotcrete

The physical properties of Gunite and later shotcrete were investigated from the very beginning. During the Gunite period, 1911-1930 the shotcrete literature describes many tests for compressive, tensile, and flexural strength, in addition to bond, permeability, shrinkage, and soundness. While the majority of the results were positive, there were disparities which could be traced to a lack of uniform standards or investigator error. Unfortunately, in most cases the experimental procedures and backup data are sketchy and unsubstantiated, so pinpointing the actual problem areas was almost impossible. However, for proper evaluation, it is essential that reproducible standards be available to and used by the industry on a regular basis.

7 Shotcrete Standards

Prior to the recent past the need for suitable shotcrete standards has been satisfied by the application of ASTM concrete standards to shotcrete technology. This has occurred primarily because the basic difference between the two technologies is the manner of placement. Hardened shotcrete whether mortar or concrete is very similar in appearance and properties to concrete and by definition may be evaluated using

concrete standards with some modification. Current ASTM standards which include test methods, specifications and practices for concrete and concrete related materials are classified under one of the following:
. Materials
. Manufacture/Production
. Plastic State Properties
. Hardened State Properties

Presently the following standards listed under these headings can be applied to shotcrete.

7.1 Materials
ASTM has standards for almost all the materials for standard or regular shotcrete applications -
Cement: (C-150;C595)
Aggregates: (C33;C330;C637) gradation should conform to table 2.1 of ACI 506R-90.
Admixtures: (C1141)
Pozzolans: (C-618);(C1240)
Reinforcing: (A615;A616;A617;A706;A767;A775;A185)
Fibers: (C-1116)
Anchor Bolts: Not available
Curing Materials: (C171;C-309)

7.2 Manufacture/Production
This division includes the batching, mixing and placement of the shotcrete and is addressed in the ACI 506.2 specification. However, ASTM C685 and C94 are referenced in the 506.2 specification and can be used with slight change in shotcrete applications.

7.3 Plastic State Shotcrete
Presently there are no concensus standards available or in use. One major problem is that undisturbed samples are difficult to obtain from plastic shotcrete and the need for plastic shotcrete standards is somewhat moot.

7.4 Hardened State Shotcrete
The shotcrete subcommittee of ASTM has tentatively reviewed the concrete standards and has found 28 usable with shotcrete. These standards include bench, field and research type tests that can be adapted or converted to shotcrete use with some modification, usually in sample preparation. Under these circumstances the existing standards for shotcrete are more than adequate for the construction industry's needs. The application of these standards to field conditions must be handled

judiciously and with care by the specifier to avoid conflict and to protect the interest of all concerned parties, especially the consumer.

8 Future Standards

We should not stand still - maintaining the status quo will only lead to retrogression. Tomorrow's standard needs will be different from those of today or yesterday. Compressive strength has always been a primary physical property for shotcrete acceptance. As the technology matures, durability, and as a consequence, service life, are becoming more important factors in both concrete and shotcrete construction. Very high strength has always been one of the more impressive properties of dry-mix shotcrete however, high compressive strength alone is not necessarily an indicator of quality shotcrete. The buzz word of today is "High Performance Shotcrete" which includes all those properties that only appropriate standards can facilitate.

What are the standards that High Performance Shotcrete will need in the years to come? In the area of <u>Materials</u>, prepackaging of shotcrete materials, both standard and proprietary mixtures, has come into vogue. In the past, prepackaging standard mixtures was primarily to provide quality shotcrete ingredients for distant and/or somewhat inaccessible locations. Proprietary shotcrete mixtures are being marketed at an increased pace to provide property enhancement of the in-place shotcrete. The physical properties and mix design vary from manufacturer to manufacturer with serious questions as to cost effectiveness. The ASTM shotcrete subcommittee has begun the process of developing a standard for prepackaging materials for shotcrete. This standard will <u>not</u> provide a procedure for establishing the relative merits of each product but hopefully will allow each consumer to make judicious choices.

In the area of <u>Shotcrete Manufacture/Production</u>, some questions have arisen concerning the efficiency of various types of equipment in producing quality shotcrete. Conventional wisdom, and some past research indicates that the term "high velocity" has great impact on the hardened properties of shotcrete. Careful, credible research on the importance of velocity or velocity of impact is needed to clearly define the concept. As a consequence standards may have to be developed to qualify various combinations of delivery and support equipment and how they operate. These standards would help determine whether certain equipment produces high velocity shotcrete as defined by ACI. ACI would continue to establish the standards for shotcrete placement techniques.

The subject of <u>Plastic State Properties</u> of shotcrete is much more difficult to address since there is little demand to create standards in this area. However, preliminary discussions in the ASTM shotcrete subcommittee have covered means and methods to determine the air content of plastic shotcrete. Procedures are available but the need has been questioned since there is some feeling that entrained air may not be absolutely necessary if the shotcrete is of high density and low permeability. One plastic standard that is being processed by the ASTM shotcrete subcommittee is "A Standard Test Method for Determination of In-Place Dry-Mix Proportions of Shotcrete." A version is currently in the ACI 506.2 specification addendum, but cannot be considered a concensus standard.

One of the major problems in concrete technology is that acceptance criteria are based on hardened state standards which are usually determined 28 days following placement. If standards are not met, correction and replacement becomes a complex problem. There is a critical need for an entirely new group of Plastic state standards which predict quality immediately, while the shotcrete is still plastic. The larger question is whether there is a body of sophisticated technology available that can achieve this goal.

As to <u>Hardened State Shotcrete</u>, the area is well covered by existing standards, however, there will always be a need for new innovative approaches to enhance quality control and assurance procedures.

9 Closure

It is necessary that our shotcrete standards be upgraded constantly to keep pace with the many advances in the technology. The creation of realistic shotcrete standards should be driven by the need to produce the very best product, not by commercial or private interests. Also, we must be careful not to over regulate by establishing a highly complex set of standards that can only lead to confusion and inequities.

25 REHABILITATION OF CONCRETE STRUCTURES IN GERMANY BY SPRAYED CONCRETE

G. Ruffert
Torkel Germany, Essen, Germany

Abstract

The spraying of concrete offers significant advantages for repair and upgrading of all kinds of concrete structures, because it needs no formwork, can be applied in thin layers with excellent bond to the substrate, and requires only small mixing and spraying equipment. The method was introduced in Germany in 1920, when the founder of Torkret - Company, Mr. Weber obtained the patent for his spraying machine. Consequently in Germany the spraying of concrete was during long years and is up to now known as „ torkretieren". The need to reconstruct a large number of concrete structures after the last war, made it necessary to establish a special standard for the use of Sprayed Concrete for this purpose. So after different „Guidelines" and „Recommendations", in 1976 the German Standard Committee published the first official standard for Sprayed Concrete, DIN 18551 [1]. To day there are several hundred construction companies in Germany, executing repair and reinforcing of concrete structures by this method.

Keywords: corrosion, rehabilitation, standards

1 Standards

With the advanced age of our concrete structure and growing environmental problems, structural repairing has become a rapidly progressing engineering branch. For many years concrete has been considered as a material which is absolutely resistant against all requirements of use and environment. The increasing number of damages, caused for the most part by insufficient steel protection, has it made necessary to set technical rules for protection and repair of concrete structures. This applies for buildings as well as for bridges.

Even as steel - corrosion, caused by insufficient concrete cover, is certainly responsible for the bigger part of all concrete deterioration, there are other causes such as

- fire
- use of anti - freezing agents
- chemical corrosion

that will make it necessary to execute constructive repairing

In 1990 the German Committee for Reinforced Concrete (D.A.f.Stb.) drew up a guideline [2] dealing with the planning and implementation of protection and repairing measures for concrete structures. According to the guideline, all protection and repairing measures must be designed and supervised by an engineer, specialised in this kind of work. His work includes the assessment of the existing structure as well for load-carrying capacity as for fire resistance.

Which are the essentials of this guideline?

The guideline sets standards for all materials used for concrete protection and repair. For the partial replacement of load-bearing concrete structures, it requires generally materials with a compression-strength >30 N/mm² and a modulus of elasticity >30 kN/mm². As this minimum requirements - absolutely necessary for load - bearing cooperation between old and new concrete - can not be obtained with polymer-modified mortars, widely used for surface repair, the replacement of loadbearing parts requires normally the use of Sprayed Concrete. The guideline itself gives no special requirements for the use of Sprayed Concrete, but refers to the German Standard DIN 18551

This standard applies to all kinds of Sprayed Concrete jobs, as well for tunnelling as for repair/upgrading of reinforced and prestressed concrete structures. Different to Sprayed Concrete standards in other countries, DIN

18551 gives not only rules for the composition, spraying and testing of Sprayed Concrete, but a great part of this standard is dealing with the special problems of loadbearing co-operation between old and new parts of the structure, giving even precise recommendations for design and stability calculation of composed structural members. This part of the standard is of special importance; as all national concrete codes and standards are aimed towards the design of new structures, posing many questions for the repairing and reinforcement of existing structures, the German standard is largely used in our neighbouring countries.

DIN 18551 applies only to concrete composed in accordance with the German Standard for Concrete and Reinforced Concrete, DIN 1045[3], that means for concrete without addition of resins or steel-fibres. As the properties of concrete will considerably be changed by this additional materials, the standard formulas for load-bearing calculation can not be applied. Sprayed Concrete with resins or fibres may thus generally not be used for structural repair or strengthening of concrete structures.

As both technical rules are considered to be official German Standards, their application is obligatory for all public works. By mention in the official Technical Terms of Trade for the Building Industry (ATV -VOB) [4], they are in fact now parts of contract for all kinds of repairing jobs

2 General Requirements

Interventions on existing structural systems may be necessary to restore or to increase the load-bearing capacity and/or stiffness of damaged elements. Principally there is no difference between repairing and structural reinforcement of loadbearing concrete members.

Apart from the special requirements for the new installed materials already mentioned, the main advantage of the use of Sprayed Concrete consists in the fact, that the excellent bond produced by surface preparation and spraying with high impact, can be used to transfer additional forces from the old to the new part of the member. Thus the German standard for Sprayed Concrete allows to introduce the excellent adherence between old and new concrete into the stability calculation for repaired or strengthened concrete members.

That such a special stability calculation will be necessary for the repair of structures badly damaged by fire is obviously out of question. But not so obvious are the material requirements, if only the concrete cover, destroyed by corrosion has to be replaced. As the obvious cause for the damage is the insufficient corrosion protection for the reinforcing bars, repairing measures are often only aimed to renew this protection and to restore the measures of the member. Thus the second main function of concrete cover requiring a sufficient thickness of the concrete layer, the introduction of tensile stresses into the steel bars is not assured. It is the same with an other key requirement, the necessary fire protection of the fire sensitive steel bars, which requires also a sufficient thickness of concrete cover in order to ensure adequate means of escape as well as access for firemen into the building.

So in many cases it may be not sufficient to restore the corrosion protection by placing a new thin layer of polymer - modified - mortar, but it will be necessary to remove the old concrete cover and to replace it by a new layer of Sprayed Concrete with the required thickness and qualities.

The application of Sprayed Concrete for repairing /strengthening of load-bearing concrete structures not only requires skilled nozzlemen, but also qualified engineers for planning and supervising this kind of structural work.

3 Requirements for Repair by Sprayed Concrete

Basically there are three problems in any rehabilitation job by Sprayed Concrete, that will determine function and future reliability of the structure:

1. Quality of Sprayed Concrete;
2. Bond in the joint between old and new concrete;
3. Load bearing co-operation between the old and new part of the structure;

The first point, quality of Sprayed Concrete is obtained by sophisticated mix design - using a maximum aggregate size of 8 mm - and by skilled spraying. In order to assure the loadbearing co-operation between old and new parts of the member , it is not only necessary to assure a minimum strength and elasticity resistance. but also to set a limit to the maximum concrete strength for the sprayed concrete. It is obvious, that the co-operation conforming to the basic rules of structural calculation is not assured, if the new concrete has a much higher strength and therefore lower ductility as the older parts of the concrete member.

The second point, strength of bond, is guaranteed by substrate preparation and the impact energy of spraying. The minimum requirements for substrate preparation are the removal of all deteriorated and disintegrated parts, the creation of a rough surface by sand-/ ore waterblasting and the removal of dust by rinsing with water under pressure. As laid down in DIN 18551, surface preparation will normally be sufficient, if tightly embedded aggregate grains are exposed. If necessary, a minimum surface tear -off strength of 1 N/mm^2 has to be shown by a pull-off test. Flame - blasting can not be recommended for surface preparation, as steel - bars and deeper layers of the structure may be unfavourably affected by the high temperatures.

The third point - load-bearing cooperation - requires in addition to concrete quality and bond strength an adequate dimensioning of the repaired/reinforced member and - in order to assure the transfer of forces - a sufficient anchorage of added reinforcing bars, conforming to the rules fixed in German Standard DIN 1045; Concrete and Reinforced Concrete.

An other important point to consider is shrinkage. It is a well known fact that Sprayed Concrete, due to small aggregate size and high cement content, is exposed to high shrinking. Other as in new structures, where the shrinking of concrete is mostly due to the lost of water by evaporation, when placing a thin concrete layer on existing members the old concrete will extract a good part of water from the new concrete by additional capillary extraction. Also their is no free shrinking possible, as the new concrete layer is thoroughly attached to the old part of the structure. In order to avoid cracks and detachment in the joint by increased shrinking in the first days, the applied layer of Sprayed Concrete must therefore thoroughly be cured by moistening (minimum 7 days). Not to recommend is the application of curing compounds, as this will make it impossible to replace the water extracted by the old concrete. In order to prevent cracking, a minimum of skin reinforcement is required if the thickness of the concrete layer exceeds 5 cm.

In order to show the essential points of DIN 18551 concerning the use of Sprayed Concrete for structural repair, there will be given some examples of jobs executed by Torkret Company, Germany.

4 References

[1] DIN 18551; Spritzbeton, Herstellung und Prüfung 3/92

[2] Richtlinie Schutz und Instandsetzung von Betonbauteilen, Deutscher Ausschuß für Stahlbeton

[3] DIN 1045, Beton und Stahlbeton, Herstellung und Bemessung, 1988

[4] ATV - DIN 18349, Betoninstandhaltungsarbeiten 12/92

26 TRAINING AND CERTIFICATION SCHEME FOR SPRAYED CONCRETE NOZZLEMEN IN THE UK

G.R. Woolley
Dept of Civil Engineering, University of Leeds, Leeds, UK
C. Barrett
Construction Industry Training Board, Kings Lynn, UK

Abstract

The Quality of Sprayed Concrete is dependant on the operatives in the field. Recognising this, The Sprayed Concrete Association has, in conjunction with the Construction Industry Training Board, introduced a formal training and certification scheme for operatives.

This Paper describes the formal training which aims to provide operatives with a knowledge of the constituent materials used, an understanding of the practice of concrete, spraying technique and methods of testing materials, and the hardened concrete. An important part of the training concerns Health and Safety at work. Details are presented of the practical test, its assessment and the subsequent examination of hardened concrete. The associated written/oral examination of candidates is also discussed.

Key words: application technique, certification scheme, materials, nozzlemen, operatives, pot-man, sprayed concrete, testing.

1 Introduction

Increasingly organisations investing in capital works or undertaking maintenance repairs of property are demanding quality assurance, particularly in respect of workmanship. It follows that high and consistent levels of skill are required from those engaged on these works. In addition, it is now being demanded by some, for Contractors to offer what is effectively a self-regulatory control of workmanship through Q.A. schemes.

In a move designed to offer a measure of control of workmanship and to eradicate what is perceived by some to be a 'Cowboy' Industry, The Sprayed Concrete Association (SCA) in 1993 introduced a Certificate of Competence for Nozzlemen. Arrangements were made with The Construction Industry Training Board (CITB) for the
testing of Sprayed Concrete Nozzlemen. An important aspect of these efforts to improve operator skill and overall competence is the provision of classroom studies by CITB on the materials, testing and techniques associated with sprayed concrete.

2 Training and study

2.1 General

In 1994 the Sprayed Concrete Association approached the C.I.T.B. with the view to develop and introduce an optional short training module. Its aim to provide the experienced nozzleman with the underpinning knowledge necessary to pass the S.C.A formal assessment, and provide a sound basis for future training and development of the individual. A three day training programme was developed broadly covering the properties of concrete, spraying concrete, Health and Safety and the COSHH [1] regulations.

The training module assumes that candidates possess the necessary practical competence, know what to do, and is structured to emphasize the reasons for and consequences of incorrect and poor practice. It is offered on the three days prior to a planned assessment of candidates for award of a Certificate of Competence. Depending on planned assessment of candidates, the programme will be biased towards the wet or dry technique as demanded. Candidates may elect to attend the training module and return for assessment at a later date, attend both training and assessment in one week, or take the assessment only.

2.2 Properties of Concrete

During this section of training, candidates are instructed on the differences between ordinary concrete and sprayed concrete, the method of placement and the numerous good practices which apply to both techniques. Constituent materials, cement, aggregate, admixtures and fibres included in sprayed concrete are reviewed. Methods of manufacture, delivery, storage, batching and mixing, physical properties, impurities and quality control are all considered. Where appropriate, measurement of physical properties are demonstrated, and the tried " craftsmans' techniques " are considered against accepted quantitive methods.

Studies of the plastic and hardened properties of concrete are developed in detail. Setting and hydration of concrete, cohesion and bleed, effect of time and temperature, all factors controlling progress are identified and discussed. The influence of a dry mix nozzleman's technique, the effect of free water/cement ratio on strength is explained and demonstrated. This practical session develops to encourage candidates to participate and consolidate the main learning points previously discussed. Durability, effect of material selection, mix design compaction and curing are demonstrated.

2.3 An introduction to sprayed concrete

This session begins with a brief history of sprayed concrete and discusses recent developments in the practice. Course participants are invited to identify the advantages and disadvantages of this method of placement, consider the two processes, discuss batching and mixing methods and review common mix proportions. Participants are invited to share their experiences by describing the type of work with which they have been involved, plant used, sequence of work, materials sprayed and problems encountered.

2.4 Preparation of surfaces

This period deals with four areas, preparation of substrate, reinforcement, formwork and surface regulation. Reasons for and methods of surface preparation are considered. Provision of reinforcement, storage, labelling, handling on site, and fixing are demonstrated. Inclusion of fibres as an alternative to reinforcement in sprayed concrete is considered. Use of formwork, its design and provision are illustrated. Methods of surface regulation with particular emphasis on cover to steel are discussed with course participants inviting their experiences from site.

2.5 Plant

Training is restricted to the gun/pump, nozzle, water pump and ancillary equipment used to connect the plant together. Operating principles of the various types of pot/pumps are described, nozzle design, operation and maintenance are discussed. Set-up procedures are demonstrated using the equipment hired for the period of training and assessment. Compressed air plays an important part in the sprayed concrete process, and the volume, pressure and moisture condition for successful spraying is discussed. Time is spent identifying the risks involved when using compressed air and the correct safety precautions to be taken are assessed and implemented.

Concrete is sprayed to demonstrate typical problems that can occur and the correct remedial measures illustrated. The likely causes of blockages are identified and the corrective/preventative measures discussed.

2.6 Techniques of application

Arguably the most important area of the training module, and where possible, instruction is in the practical form. Classroom instruction considers accepted good practice, including stance, spray pattern, optimum spraying distance, communication with the Pot-man, spraying sequence, encapsulation of reinforcement and evacuation of rebound. Course participants are invited to spray test panels. Assessment candidates become familiar with the equipment to be used for their examination, and individual coaching is possible. Sprayed panels are dismantled, allowing inspection and examination of the compacted concrete, defects are highlighted and discussed.

2.7 Testing and control
Testing of the plastic and hardened concrete is discussed. Mould design, age of testing, effects of curing and locations of samples are considered. A typical testing laboratory is visited where brief demonstrations of sample preparation, storage and testing are given.

2.8 Health and Safety
The teaching of Health and Safety in relation to sprayed concrete plays an important part in the overall training programme. It is a taught subject in the programme, but remains a constant theme through all lessons and practical demonstrations. Throughout the course, relevant safety issues are highlighted and discussed. Any candidate spraying concrete during the training module, or assessment, failing to comply with the regulations, is immediately stopped from spraying and risks failing the assessment.

3 Certification examination

3.1 General
The examination of Nozzlemen is sub-divided into two distinct parts : an oral or written test of knowledge on matters relating directly to the practice of sprayed concrete and the spraying thereof, and a practical demonstration of spraying technique. Emphasis on technique is reflected in the marking scheme which awards 70% of total marks for technique and subsequent adequacy of the hardened concrete. Candidates are judged by their ability to produce a dense, impermeable concrete, fully encasing reinforcing steel with a closed, even surface [2]. To be awarded the Certificate of Competence, a candidate has to achieve a minimum of 70% marks overall with no less than 60% in both sections of the examination. The examination is conducted by an Independent Engineer who makes recommendations to CITB for the award or otherwise of the Certification.

3.2 Oral/written test
A series of questions have been devised which are designed to examine the knowledge of candidates in their respective specialities namely: dry mix spraying, wet mix spraying, structural, underground, refractory and pre-stressed tank sprayed concrete. All candidates have to answer questions on general topics associated with sprayed concrete, and also on health and safety. Up to 39 questions are available to the certification examiner in each specialist subject with around 25 on general or safety matters. All these questions are published in the Association's Guide to Certification [3]. Candidates are expected to have studied the questions and answers which are in the simple yes/no format. In addition there are a number of supplementary questions, not openly disclosed to candidates, available only to the certification examiner. The latter are used, as necessary, to satisfy the certification examiner that the candidate actually knows his subject or if the correct answers have been learned 'parrot fashion'. Questions are posed in random fashion rather than in sequence again to reassure the examiner. Typically a minimum of 15

questions shall be answered on each subject making a total of not less than 60. In reality the examiner will ask in the region of 100 questions or more, as he seeks to determine a satisfactory standard of competence. A total of 30% of marks are available for this section of the examination.

3.3 Workmanship demonstration

A candidate has to demonstrate his ability to correctly and successfully apply the type of sprayed concrete for which he is seeking a Certificate of Competence. A site mixed cement and aggregate is used for the test, additives or admixtures are not allowed. As a supplement to the test a candidate may elect to spray a proprietary pre-bagged material or site blended mixture in addition to the site mixed sprayed material. Mix proportions are selected which will produce a 28 day compressive strength of at least 40 N/sq. mm.

Tests are conducted using equipment normally used for the type of sprayed concrete for which the candidate is being examined. A competent crew, including an experienced pot-man is provided to assist the candidate in the demonstration. A curing agent or plastic sheeting is available to the candidate for the proper curing of his demonstration panels.

3.4 Test panels

For the dry mix spraying of structural concrete, candidates are required to fill three timber panels. These are illustrated in Figure 1. The open panel has to be filled and left with a trowelled finish, the reinforced panel has to be filled and left with a cut and flash finish. The smaller panel has to be filled and cut back for use to extract cores for the determination of estimated in-situ cube strength.

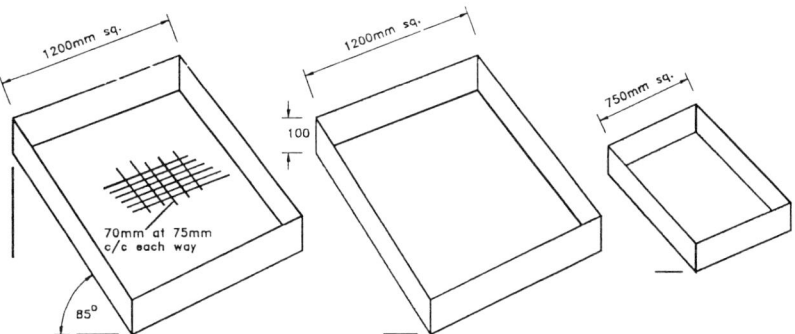

Panel 1 - Cut & flash finish   Panel 2 - Trowelled finish   Panel 3 - estimated in-situ cube strength

Figure 1. Dry mix sprayed structural concrete demonstration of technique panels.

For dry mix spraying of underground sprayed concrete two panels are required to be filled. Size of panels, orientation and inclusion of reinforcement are given in Figures 2 and 3. Panels are to be filled flush, but no finishing work is required after filling. Cores are extracted from both panels to determine estimated in-situ cube strength.

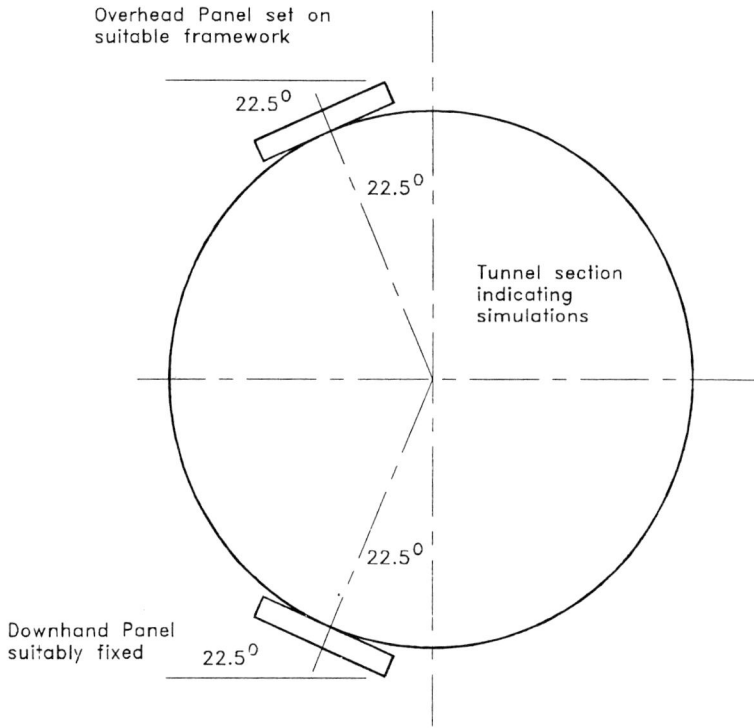

Figure 2. Underground sprayed concrete - test panel formation.

At this time panels and practices are being developed for the wet mix spraying of structural and underground sprayed concrete. Further work is required to develop examination methods for refractory and prestressed tank sprayed concrete.

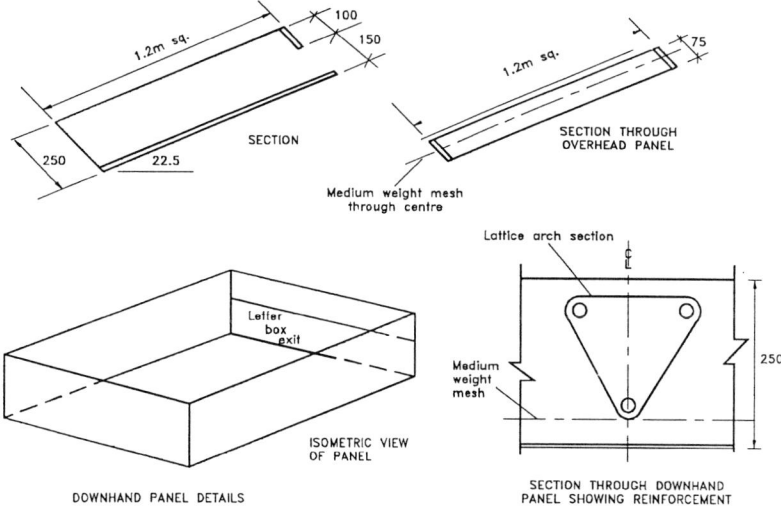

Figure 3. Underground sprayed concrete - Panel Details.

3.5 Marking

Marking of the oral/written examination is quite straight forward. Marks are given for correct answers and the percentage marks computed and added to the examination total.

Marking for the practical examination is more difficult. The total 70 marks have been sub-divided into 13 sections (12 for underground sprayed concrete), typically a section earning around 3/5 marks. To reflect the importance of two particular aspects of spraying, spray pattern and inclusion of rebar carry 15 marks each. The certification examiner is thus able to assess overall quality of workmanship, but at the same time is able to penalise for poor technique. Sub-division of marking schemes are presented in Appendix 1. Throughout the practical demonstration the certification examiner records in contemporaneous notes all features of a candidates actions during the test, notes which form the basis of a formal report provided to CITB.

Examination of the hardened concrete is an important part of the examination. Shutters are stripped and any defects are noted. Cores are taken from the concrete to determine estimated in-situ cube strength and to allow examination for defects within the concrete or behind rebars. Marks may be deducted for defects found at this stage.

To complete the examination procedure, the certification examiner prepares a detailed report on each candidate's practical demonstration and their hardened concrete test panels. A recommendation is made to CITB for the issue or otherwise of a Certificate of Competence.

4 Conclusion

Experience has shown that testing of a candidates knowledge is better given orally. This allows the certification examiner to help a candidate overcome ' examination nerves ' and satisfy himself as to depth of knowledge of a candidate.

Procedures established by SCA and developed by CITB for the examination and testing of nozzlemen, using the dry mix process for structural and underground sprayed concrete have been shown to be effective. Developments are on-going to extend the range of tests to include all specialisms and encompass wet mix spraying.

Not all persons taking the classroom course of study provided by CITB elect to undertake the certification examination. Currently around 65% of candidates are successful when taking the examination.

Provision of a training programme and certification procedure with examination by an independent examiner, gives the sprayed concrete industry an opportunity to improve quality of workmanship and offer a quality control hitherto not always thought available by potential clients.

Acknowledgements

The authors thank both SCA and CITB for permission to prepare and present this paper.

References

1. The Control of Substances Hazardous to Health Regulations, 1994, Her Majesty's Stationery Office, London.
2. The Concrete Society, Concrete Society Code of Practice, Concrete Society, Slough, U.K., 1980
3. The Sprayed Concrete Association, Guide to Certification of Sprayed Concrete Nozzlemen, Sprayed Concrete Association, Aldershot, U.K., 1993

Appendix No. 1

Sub-division of marks awarded for the practical demonstration of dry mix process sprayed concrete

	Structural	Underground
Personal protection (H & S)	5	5
Instructions to Pot-man	5	5
Wetting sub-strate	3	3
In-filling corners- incl. build up of rebound	3	3
30 deg. top surface	3	3
Nozzle control	2	6
Spray pattern	15	15
Filling behind rebar	15	15
Too wet / dry	2	2
Over / under filling	5	5
Evenness of filling	4	4
Flash coat	4	--
Visual appearance	--	4
Trowelled surface	4	--
Marks available in practical test	70	70
add oral /written questions	30	30
TOTAL MARKS AVAILABLE	**100**	**100**

PART SEVEN
SPRAY PROCESS

27 PARTICLE KINEMATICS IN DRY-MIX SHOTCRETE – RESEARCH IN PROGRESS

H.S. Armelin, N. Banthia
Civil Engineering Dept, University of British Columbia, Vancouver, Canada
D.R. Morgan
AGRA Earth & Environmental Ltd, Burnaby, Canada

Abstract
Despite its widespread application throughout the world, dry-mix shotcrete has as one of its main drawbacks a high rebound rate, which can lead to losses of up to 50% for the aggregate phase and 70% for the steel fibres. In this preliminary study, kinematic values of particle velocity and acceleration were evaluated for individual aggregate and steel fibre particles using a high-speed camera capable of up to 1000 frames per second. The influence of variables of air volume, nozzle type, shooting direction, aggregate size and steel-fibre geometry on particle velocity and acceleration were also evaluated.

Results show a relationship between aggregate particle size and cruising velocity that agrees with the well known dependence of rebound on particle size, with larger aggregates tending to rebound more. Overall, the characterization of shotcrete kinematics yields valuable information that should lead to the development of a rational theory for a better understanding of the fundamental mechanisms of particle rebound.
Keywords: aggregate velocity, fibre velocity, rebound, nozzle type

1 Introduction

Due to the fact that it is pneumatically applied, shotcrete is characterized by high particle velocities with the placing and compacting being done essentially in an impact process of aggregate and paste striking the shooting surface. As a result, after impact, some particles fail to attain stability and fall to the ground in the form of rebound.

Although rebound losses are usually below 15% for the wet-mix shotcrete process, for dry-mix shotcrete rebound is usually in the range of 20 to 40% [1] with values as high as 50% being reported [2].

Besides the obvious implications that rebound has on the cost per cubic meter of in-place shotcrete, its consequences are actually more far reaching. Because it is primarily composed of aggregates, rebound tends to cause the in situ material to have a cement content significantly higher than the design mix, with figures reported for the in situ cement content usually showing in excess of 500 kg/m^3 [3]. These cement rich mixes are more prone to cracking caused by thermal or moisture gradients and more vulnerable to shrinkage cracking and the problems this can lead to in some aggressive exposure environments.

In the case of steel-fibre reinforced dry-mix shotcrete, the rebound problem is even more severe, with reports of up to 70 to 80% of the steel-fibres being lost in the form of rebound [4][5], leading to diminished post-cracking reinforcing ability of the in situ shotcrete.

However, despite the fact that rebound is one of the major concerns of the dry-mix shotcrete industry today, knowledge on how to minimize it is still empirically based and a rational theory of particle rebound has not been developed to date.

Thus, some basic questions concerning the problem of rebound in dry-mix shotcrete remain unanswered:

- Why does the wet mix process show a 10 to 15% rebound while in dry-mix shotcrete rebound of up to 50% is reported ?
- Why do large aggregates rebound up to four times more than the smaller sizes (70% for the 9.5 mm aggregate compared to 15% for the 0.075 mm material) ?
- Why is steel-fibre rebound so high ? Are there fibre geometries that rebound less ?
- Do particles accelerate or decelerate in the shotcrete stream ?

In order to develop a general theory of particle rebound for shotcrete, one has to start by recognizing that the process through which particles impinge on the shooting surface and either become stable or rebound is essentially an impact process. Therefore, one of the main determining factors in the process is the particle velocity of impact.

However, characterizing particle velocities in shotcrete can be very difficult since the pneumatic process generates speeds that are impossible to be assessed by naked eye or conventional methods. As a result, despite the fact that some studies have been made using high speed photography [6][7], a state-of-the-art report on particle velocity in shotcrete [8] shows that the speed at which aggregates tend to travel is still largely unknown, with reports varying widely between 10 and more than 100 m/s (36 to more than 360 km/h).

2 Objective

The main objective of this study is to provide an interim report on the activities presently in progress as part of a more comprehensive research program being carried out at the University of British Columbia (UBC) that aims at understanding the mechanism of particle rebound in dry-mix shotcrete. As a first step in developing a general theory of particle rebound, this research program has concentrated on characterizing particle velocities (aggregates and fibres) and the main influencing parameters (nozzle type, direction of shooting, particle size and shape).

3 Materials and Method

In order to record particle velocities, a high-speed camera (model EKTAPRO 1000) was used. This camera is capable of recording up to 1000 frames per second and reports the time in milliseconds on a screen along with the moving images recorded. In order to determine particle positions, all shooting was done using a 50 mm square mesh grid as a background. Average velocity was calculated using the times of particle entrance and exit over a 300 mm wide window. The average acceleration was calculated from the change in velocity from one 50 mm wide window to the next. All velocity measurements were made over a window located 1000 to 1300 mm away from the end of the nozzle.

Because the actual shotcrete environment is too dusty for the delicate photographic equipment and because the high shutter rates used (1000 frames per second) require intense lighting, aggregate particles and fibres were shot using a single particle shooting apparatus. This equipment is essentially a compressed air gun (nozzle) that allows one to shoot as many particles at a time as desired (Fig. 1).

Two different nozzle geometries were tested (types A and B - Fig. 2) with particles being shot on both overhead and vertical wall positions. Aggregates were sieved to separate the different sizes in the standard sieve set. The aggregate particle size was also measured on screen using a pair of calipers and scaled to actual size to determine the particle size vs velocity relationship. With respect to how representative of actual shotcrete conditions the single particle shooting system is, further studies of aggregate velocity from an industrial scale shotcrete equipment show that the velocities obtained using nozzle B are close to those generated by actual dry-mix shotcrete conditions.

4 Results and Discussion

Aggregate Velocity vs Size

Results for aggregate velocity and its dependence on aggregate size for nozzle A, wall shooting are shown in Fig. 3a. These data indicate a strong relationship between aggregate size and velocity, with larger particles traveling at a speed of approximately 10 km/h while for the smaller aggregates the average speed is in excess of 20 km/h (same air supply). The same dependence of aggregate velocity on size is found for nozzle B (Fig. 3b).

It is interesting to note that the relationship found between particle size and velocity is similar to the dependence of rebound on particle size that is commonly found for

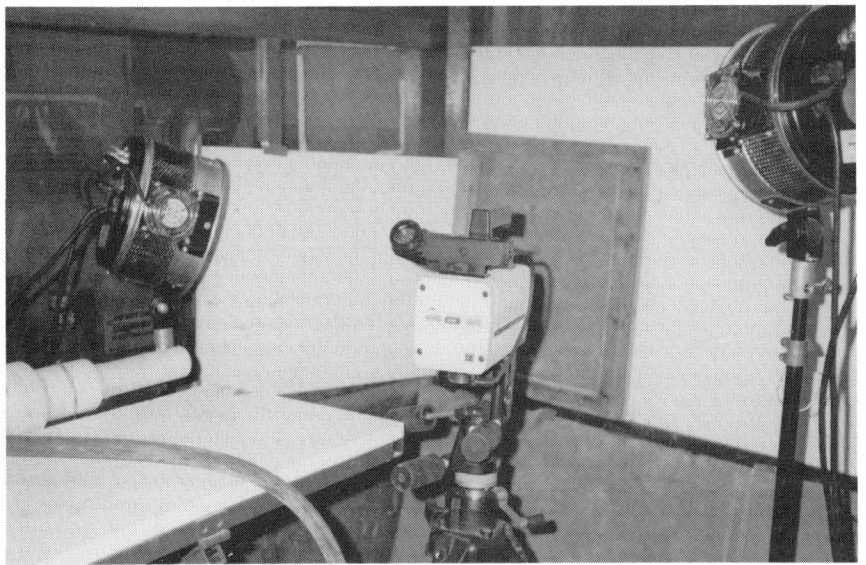

Fig. 1 - Single particle shooting apparatus used to shoot fibres and aggregates.

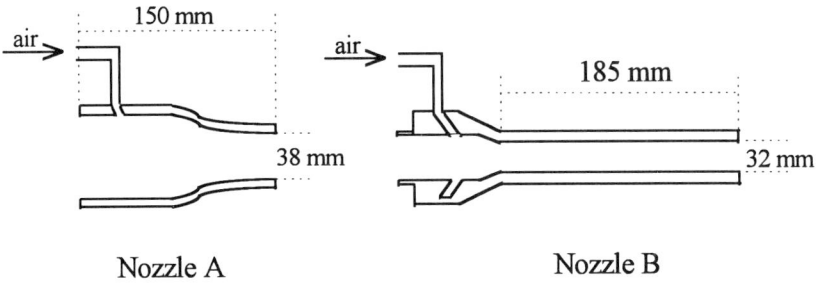

Fig. 2 - Two different nozzle geometries used to shoot aggregates and fibres.

dry-mix shotcrete (Fig. 4). Using a model of impact presently being developed at UBC it is possible to show that the low velocity of impact is one of the reasons why large aggregates tend to rebound up to four times more than the smaller size aggregates.

Aggregate Velocity vs Nozzle Geometry and Direction of Shooting

Aggregate velocity vs size profiles are presented in Figs. 3a and b for two different nozzle geometries under the same air supply and shooting direction (wall shooting). A comparison of these two figures reveals a distinct influence of the nozzle geometry on the aggregate velocities, with the longer, smaller diameter nozzle leading to velocities, on average, four times greater than the shorter, larger diameter case.

In Figs 3 b and c, velocity vs aggregate size profiles are presented for a same nozzle geometry under conditions of wall and overhead shooting. Although not as marked as the influence of nozzle shape, the direction of shooting also shows some degree of influence on aggregate velocities with overhead shooting leading to lower velocities of impact.

It is interesting to note that the dependence of particle velocity on aggregate dimensions is confirmed for all three situations tested (Figs. 3a, b and c).

As for acceleration measurements, in all fifteen situations of aggregate size, nozzle type and position of shooting tested, the average change in velocity from one 50 mm wide window to the next was found to be close to zero, indicating that virtually no acceleration occurs after the aggregate has left the nozzle and within the window size sampled. This was confirmed even for the case of overhead shooting, for which no average deceleration was found at a distance of 1000 to 1300 mm from the nozzle (not necessarily true for greater distances).

Fiber Velocity vs Shape

Results of average fibre velocity recorded for three different fibre geometries tested are shown in Table 1 for situations of wall and overhead shooting using the same nozzle and air supply.

Banthia et. al. [4] reported a higher fibre rebound for fibre geometries of high surface area and attributed this to the greater velocity that foil-like geometries tend to develop. However, despite the widely different geometries tested in this study (flat and cylindrical), given the large variability in the fibre velocities recorded (coefficient of variation between 30 and 50%), it must be concluded that, while there are some trends, no significant influence of fibre shape or position of shooting on fibre velocity was found in this study.

Fiber geometry and its influence on fibre rebound of "real" dry-mix shotcrete are the subject of further studies presently being carried out at UBC and, although the surface area is not a general parameter determining fibre velocity, for a given fibre shape (eg.,

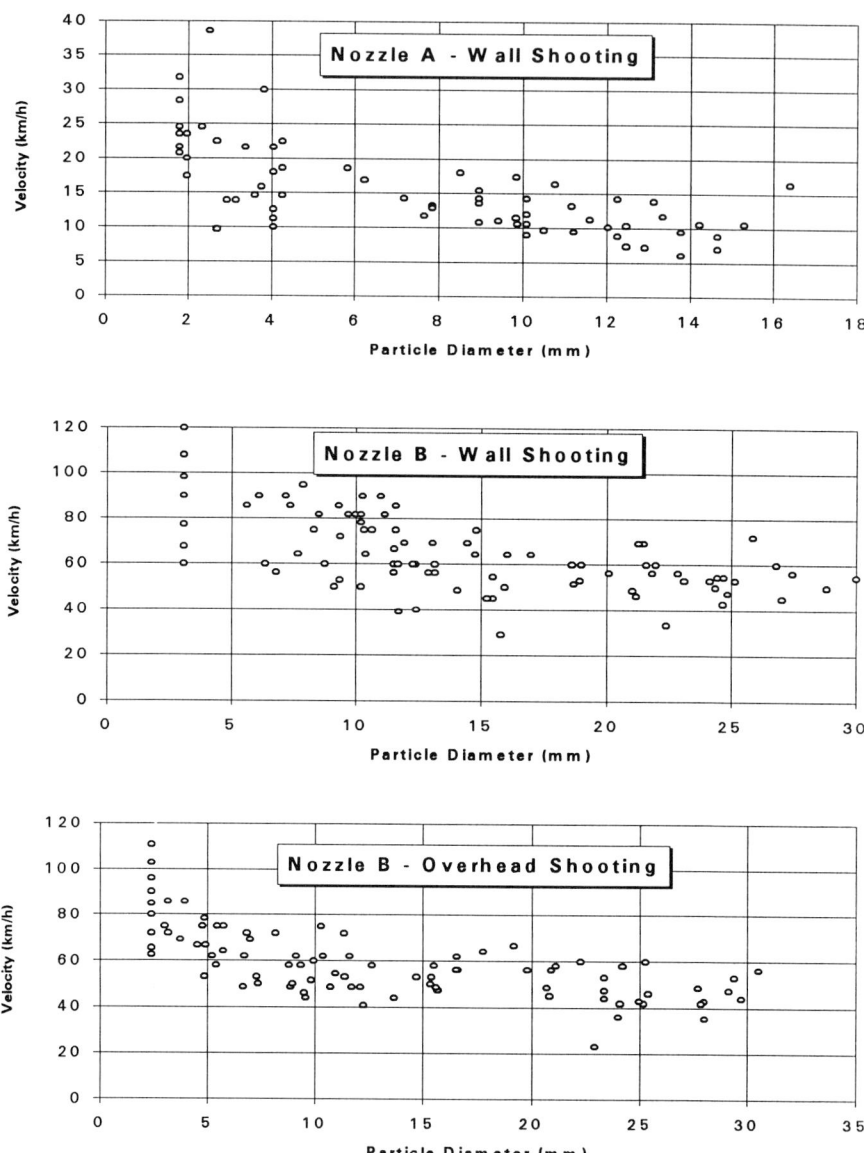

Figs 3a, b and c - Dependence of aggregate velocity on size for nozzle A (a), nozzle B for wall shooting (b) and nozzle B for overhead shooting (c).

cylindrical) preliminary results do indicate some influence of the projected fibre area in determining the fibre velocity.

Table 1 - Average velocities recorded for the different fibre geometries tested.

Fiber Geometry	Dimensions*	Nozzle B - Wall**	Nozzle B -Overhead
Hooked	d = 0.50, l = 30	58.1 / 19.1	43.3 / 24.3
Flat	t = 0.40, w = 2.4 , l = 30	61.1 / 21.7	57.8 / 19.6
Straight	d = 0.40 , l = 25.4	47.0 / 19.2	52.3 / 26.0

* diameter, length, thickness and width all in millimeters
** average fibre velocity (km/h) / standard deviation (km/h) - sample size = 20 readings

5 Conclusions and Future Activities

In this preliminary study, a technique has been developed to characterize particle velocity and acceleration in shotcrete, allowing measurement of aggregate and fibre speeds in different situations of particle size, nozzle geometry and position of shooting. The most important conclusions to be extracted from this study concern the relationship found between aggregate size and velocity as well as nozzle geometry and particle velocity.

At present, a study of particle velocities is being carried out at UBC using full scale industrial size dry-mix shotcrete equipment. Variables being tested include the influence of aggregate particle size, air flow, position of shooting and hose diameter.

In addition, a comprehensive experimental study of dry-mix shotcrete rebound is being conducted using test panels sprayed inside a closed chamber (Fig. 5) under various mix designs and different shooting techniques. The in situ and rebound material are being fully analyzed in order to determine the influence of rebound on the properties of the final shotcrete product.

Finally, a model is being developed which uses the velocity profiles experimentally found to analyze the impact event of an aggregate or fibre impinging against the shooting surface and eventually rebounding. It is hoped that this development of a fundamental understanding of the shooting process will lead to an optimization of dry-mix shotcrete with respect to minimizing rebound losses.

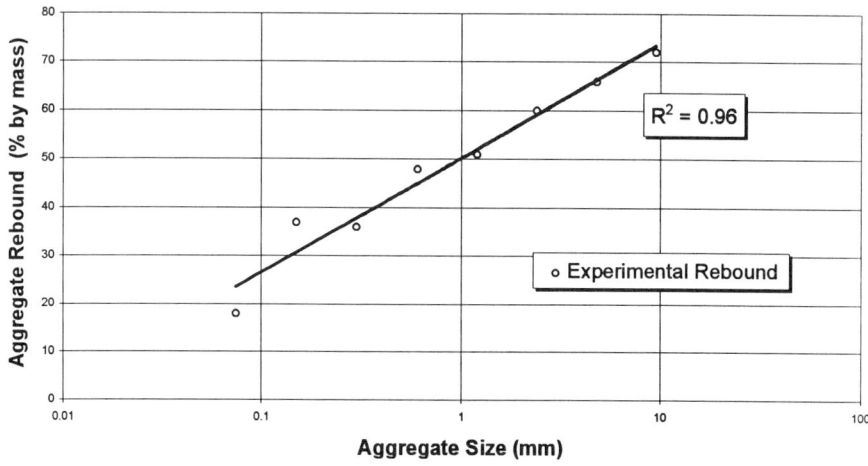

Fig. 4 - Variation in aggregate rebound with size for dry-mix shotcrete.

Fig. 5 - Dry-mix shotcrete equipment and rebound chamber presently in use at UBC.

Acknowledgments

This study was funded by the Canadian Natural Sciences and Engineering Research Council (NSERC) and supported by ALIVA Limited of Switzerland. The authors would also like to thank Dr. S. Mindess from the UBC Civil Engineering Department whose high speed camera was used for this study. Hugo Armelin was funded by The Brazilian Federal Research Council (CNPq).

6 References

1. Wolsiefer, J. and Morgan, D. R., Silica Fume in Shotcrete, Concrete International, V. 15, N. 4, 1993, pp. 34-39.

2. Warner, J., Understanding Shotcrete - The Fundamentals. Concrete International, May and June 1995, Vol. 17 Nos. 5 and 6, pp. 59-64 and pp. 37-41.

3. Austin, S. A., (Shotcrete) Production and Installation. Sprayed Concrete: Properties, Design and Application (Edited by S. A. Austin and P. J. Robins). McGraw-Hill, New York, 1995, pp. 41.

4. Banthia, N., Trottier, J-F., Wood, D. and Beaupre, D., Influence of Fiber Geometry in Steel Fiber Reinforced Dry-mix Shotcrete. Concrete International, V. 14, N. 2, 1992, pp. 24-28.

5. Armelin, H. S. and Helene, P., Physical and Mechanical Properties of Steel Fiber Reinforced Dry-mix Shotcrete. ACI Materials Journal, V. 92, N. 3, May-June 1995, pp. 258-267.

6. Parker, H. W., Field Oriented Investigation of Conventional and Experimental Shotcrete for Tunnels. Ph.D. Thesis, University of Illinois at Urbana-Champaign, USA, 1976, 630 pp.

7. Ward, W. H. and Hills, D. L., Sprayed Concrete - Tunnel Support Requirements and the Dry-mix Process. Shotcrete for Ground Support (SP-54), ACI, Detroit, 1977, pp. 475-532.

8. Glassgold, I. L., Shotcrete Durability: An Evaluation. Concrete International, V. 11, N. 8, 1989, pp. 78-85.

28 FUNDAMENTALS OF WET-MIX SHOTCRETE

D. Beaupre
Laval University, Quebec, Canada
S. Mindess
University of British Columbia, Vancouver, Canada

Abstract
This paper presents the results of a study undertaken to obtain a fundamental understanding of the shooting process. A model based on rheological behavior was developed to predict pumpability and shootability. Flow resistance and viscosity were used to represent the rheological behavior of fresh shotcrete. Important fundamental relationships were obtained between rheological properties and pumping pressure, build-up thickness and compaction of shotcrete. With a new parameter, the fresh concrete ageing rate, these relationships can be used in a model which predicts pumpability and shootability.
Keywords: build-up thickness, compaction, flow resistance, pumpability, rheology, shotcrete, viscosity, wet-mix, workability box, shootability

1 Rheology

1.1 Bingham model
Rheology is defined as the science of deformation and flow of matter. In terms of fresh concrete, rheology is related to the flow properties (the mobility) of concrete. When a shear stress is applied to a liquid, the liquid deforms and keeps deforming until the stress is relieved. Different relationships between applied shear stress and flow characteristics can be obtained. Newtonian behavior is the simplest one for a fluid: the rate of shear is proportional to the applied shear stress. Figure 1a shows the graphic representation of a Newtonian fluid, while Figure 1b represents the Bingham model which is more appropriate for fresh concrete. For the Bingham behavior, two parameters are needed to fully describe the fluid behavior: the yield value (τ_0 in Pa) and the plastic viscosity (μ in Pa.s).

Typical results obtained on different concretes are shown in Figure 2 [1]. In this Figure, the flow curves strongly suggest a Bingham behavior. The rheological behavior can be expressed in term of two parameters by the following equation:

$$T = g + h\,N \qquad (1)$$

where T is the torque to drive an impeller (Nm), g is the flow resistance (Nm), h is the torque viscosity (Nm.s) and N is the impeller angular speed (rev/s). Equation (1) is very similar to the equation in Figure 1b, which suggests that the flow resistance is related to the yield and that the torque viscosity is related to the plastic viscosity.

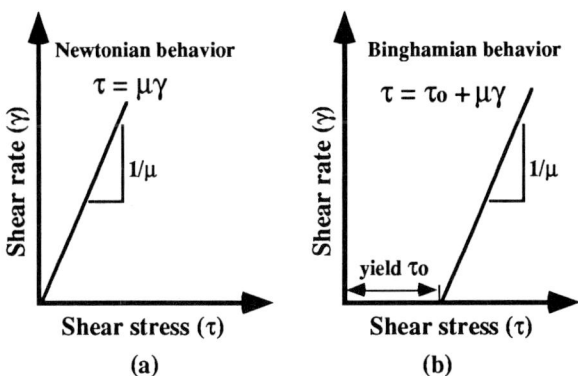

Fig. 1: Representation of the Newton (a) and the Bingham (b) model

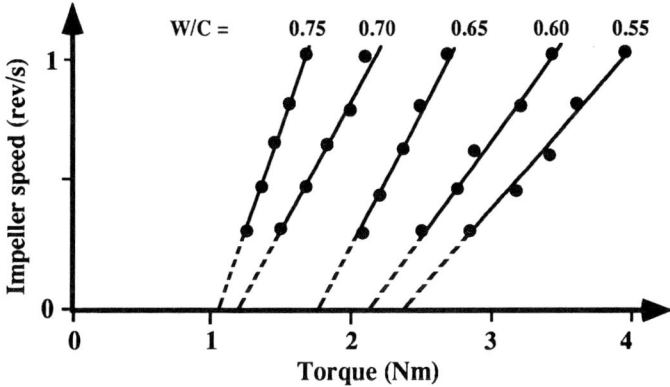

Fig. 2: Typical results from the MKII rheometer (from ref. [1])

Unfortunately, these parameters (g and h) are not in the fundamental units of yield (t_o = Pa) or plastic viscosity (μ = Pa.s). The g (Nm) and h (Nm.s) values are affected by the geometry of the apparatus with which they are measured. However, it is possible, by proper calibration, to convert them into the fundamental units for μ and t_o. Even if not in fundamental units, the rheological parameters g and h can be used to evaluate the concrete mobility or other related properties.

1.2 Ageing

The age of concrete (the time elapsed after mixing) is an important factor to consider when determining the rheological properties. Because chemical reactions occur even during the dormant period, the viscosity and the yield of the concrete change with time. Studies have shown that it is mainly the yield (the flow resistance) that changes with time [2,3]. The viscosity is not much affected by the ageing process.

To evaluate ageing, one may define a fresh concrete ageing rate (FCAR) which is the rate of change in flow resistance with time. Mixtures which are ageing slowly will have a small FCAR: they will remain workable for a long period without significant changes in rheological properties. Figure 3 shows the behavior of such a stable mixture: there is only a very small increase in flow resistance and no change in viscosity over a two hour period (*Cast 15 min* refers to the properties of the concrete 15 minutes after casting). The flow resistance of this mixture as a function of time is shown in Figure 4 where the slope of the line (0.2 Nm/h) represents the fresh concrete ageing rate (FCAR). A very low FCAR (as in set retarded concrete) is not desirable in shotcrete applications because it can over-extend the waiting period between two successive applications. Because the slump is related to the flow resistance [4], the FCAR can be used to study slump loss with time.

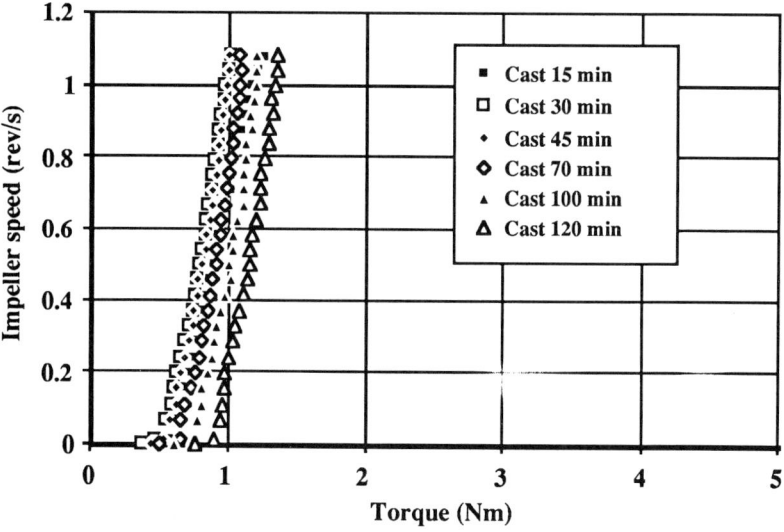

Fig. 3: Rheological test results on the same concrete at different times

In some mixtures, the flow resistance rapidly increases with time, due to a rapid stiffening of the mixture. Such a behavior is good for shotcreting applications because the waiting time between successive applications (e.g., for thick overhead applications) is reduced. Of course, if the fresh concrete ageing rate is too high, pump blockage may occur. Because the value of the fresh concrete ageing rate affects the shotcrete application, it should be taken into account in a model which predicts shootability.

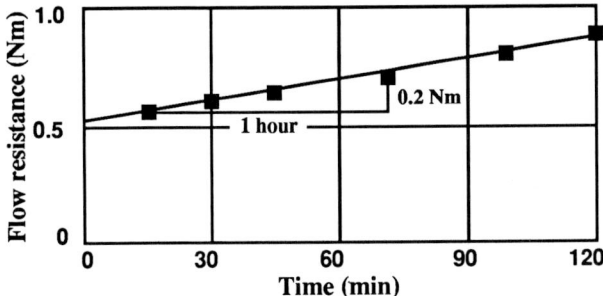

Fig. 4: Determination of the fresh concrete ageing rate (from the test results in Figure 3)

Another time dependant effect is thixotropic behavior. At rest, certain fresh concrete mixtures may exhibit a rapid increase in flow resistance. If this stiffening can be annulled by mixing, the material is thixotropic if, at rest, it become stiffer again. Two successive tests are needed to determine whether a material is thixotropic. Figure 5a shows a first test carried out on a material which exhibits a decrease of viscosity with an increase in the shear rate. Figures 5b and 5c are the possible results of a second test carried out after a short period of time on the same material. In case (b), the second up-curve and down-curve are identical to the first down-curve; the material is not thixotropic. In case (c), the second up-curve is close to or identical to the first up-curve, the material is then thixotropic.

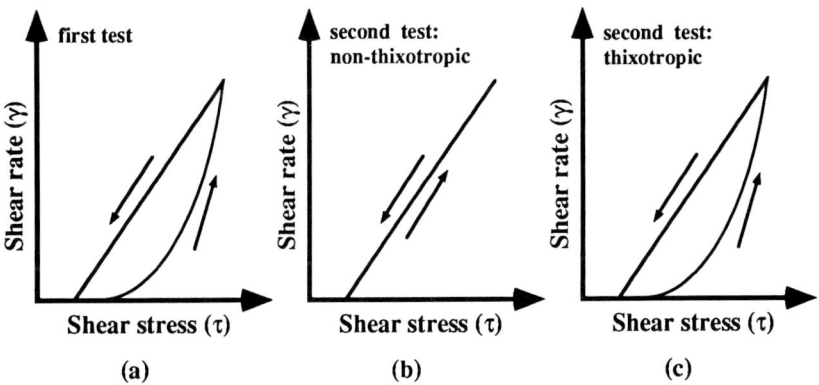

Fig. 5: Illustration of the thixotropic behavior

1.3 Effect of composition and description of the workability box

Almost any change in mixture composition may affect the rheological behavior of concrete:

- content, shape, gradation, porosity, and texture of aggregates
- content and type of cement
- presence of other cementitious materials (fly ash, silica fume...)
- use of admixtures (superplasticizers, air-entraining agents, accelerators, retarders...)
- presence of fibers (type and quantity)

- proportions of all constituents (W/C, etc.)

Interactions between constituents complicate the situation because they are not independent of each other in their effects. It has been shown (Figure 2) that an increase in W/C produces a reduction in both the plastic viscosity and the flow resistance. Figure 6 illustrates the effect of mixture composition on the rheological properties. This type of graph is very useful to visualize the effect of different parameters.

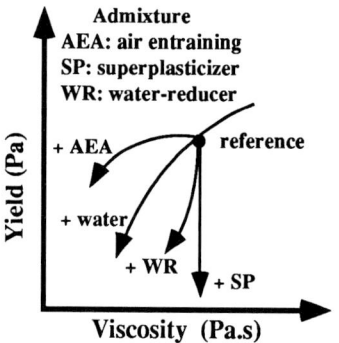

Fig. 6: Effect of changes in mixture composition on rheological properties

It has been stated that, for any given concrete placing method (for example: filling a pipe with flowing concrete), it should be possible identify a region, a workability box, which would enclose all combinations of g and h suitable for that application [1]. In Figure 7, all concretes represented by black dots (inside the workability box) would be suitable for this application while those represented by white dots would not.

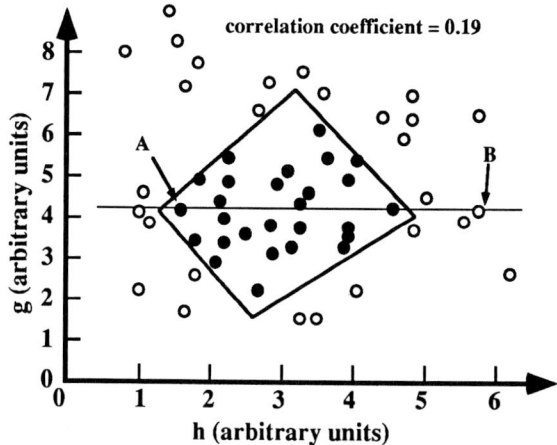

Fig. 7: Relationship between flow resistance (g) and viscosity (h), and illustration of the workability box (adapted from [1])

Concretes A and B in Figure 7 can be used to illustrate the use of the workability box. Both have the same g, and, according to Scullion, they should have approximately the same slump. However A is suitable, but not B which has a high viscosity and is outside the box. This shows the importance of considering both rheological parameters to assess workability. For certain applications such as pumpability for instance, where the viscosity is an important factor, the slump test is not sufficient to assess workability.

2 Pumpability

Pumpability, like the other flow related properties, can be estimated with the use of the two fundamental rheological properties: viscosity and flow resistance. As shown in Figure 8a which was obtained form the results of tests, an estimate of the required pumping pressure (for a given pumping system) can be obtained and used to define a pumpability box (Figure 8b) [5]. All mixtures inside the box were observed to be pumpable (i.e. no pump blockage occurred). In this particular case the results of the tests indicated that an increase in the value of g or h produced an increase in the required pumping pressure. However, in addition to the resistance due to high values of g and h, the possibility of blockage due to segregation or bleeding under pressure also exists and is possibly caused by a too low viscosity [6]. Ageing effects on the prediction of pumpability should also be considered [5].

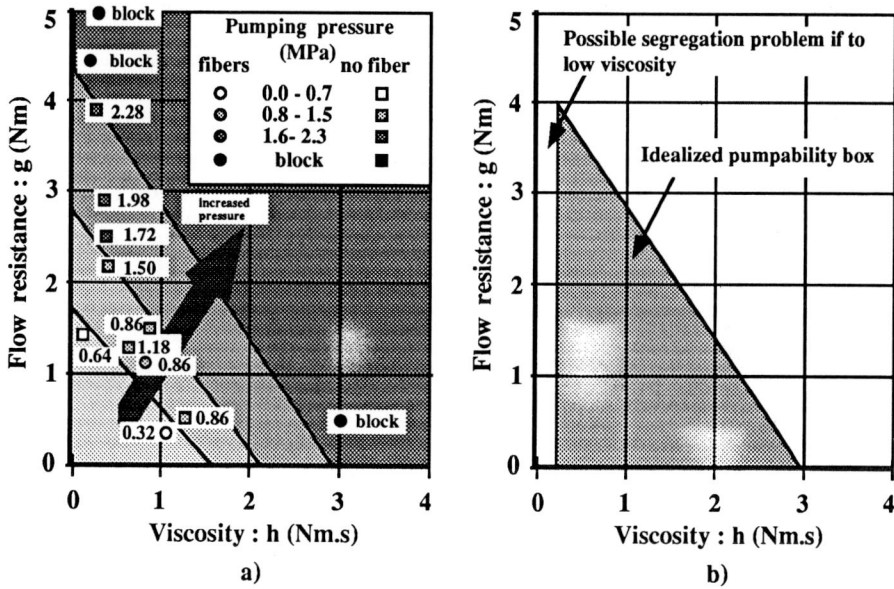

Fig. 8: Relationship between pumping pressure and rheological properties: a) experimental data (from ref. [5]) and b) idealized pumpability box

An understanding of the influence of the flow properties of concrete is probably not enough to determine precisely the pumpability. Casual observation of pumped concrete is sufficient to realize that concrete moves into the pipeline as a solid plug (Figure 9). The characteristics of the lubricating layer are therefore yet another item to consider in pumpability studies.

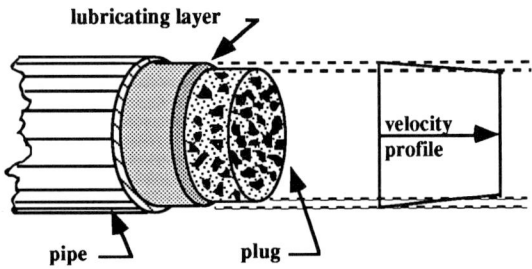

Fig. 9: Concrete in pipeline: plug flow (Browne and Bamforth, 1977)

3 Shootability

3.1 Relationship between shootability and rheology

Shootability can be defined as the efficiency with which a concrete sticks to the receiving surface and to itself. In this paper, the build-up thickness is used as a measure to assess shootability: a mixture that can be built-up to a greater thickness in a single pass without sloughing will be referred to as having a better shootability. In certain cases where thickness is of less concern, the amount of rebound may be critical and thus be used to assess the shootability of the mixture.

Up to now, there have been fery few studies on the rheology of wet-mix shotcrete. A rheology-based explanation as to why shotcrete stays in place after shooting as been proposed [7]:

> *The existence of a yield value seems to provide a good explanation as to why shotcrete is shootable. Intuitively, the higher the yield stress, the better the shootability (i.e. the greater the thickness that can be built up without sloughing). In fact, a material with no yield value (such as water) could not remain in place after shooting. Similarly, a flowing concrete with low yield value would not be suitable for shotcreting; it would simply slough off the receiving surface unless special agents (such as certain accelerators) were added at the nozzle. On the other hand, mixtures with very high yield value (low workability) could be unsuitable for shotcreting, because of pumping and consolidation difficulties.*

In fact, a good relationship between the flow resistance (g) and the maximum horizontal build-up thickness has been found [5]. Figure 10 shows that when the in-place flow resistance of shotcrete is high, the build-up thickness is high and vice versa. Based on the relationship in Figure 10, it is possible to determine a workability box for shotcrete mixtures that is suitable for a given expected build-up thickness (see Figure 11). In this Figure, all mixtures with an in-place flow resistance over the dotted line (g = 2 Nm) would satisfy the shootability requirements for a build-up thickness of 140 mm on the wall (see the dotted line in Figure 10 which indicates that a build-up thickness of 140 mm corresponds to a flow resistance of 2 Nm). In this case, as expected and as can be seen in Figure 11, viscosity was not observed to have a very significant influence.

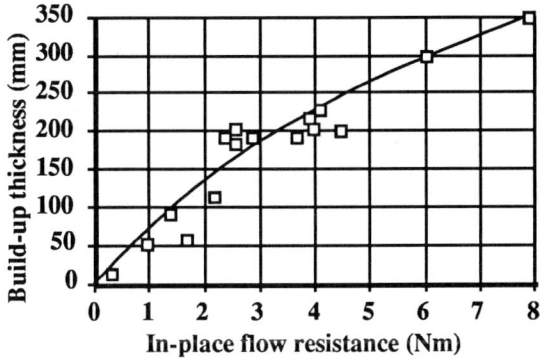

Fig. 10: Relationship between the build-up thickness and the in-place flow resistance (addapted from [5])

3.2 Effect of compaction on rheological properties and shootability

Researchers have explained the effect of compaction on the rheological properties: the impact of the mixture on the receiving surface causes compaction (expulsion of air) which produces a sudden increase of the flow resistance of the in-place material [2]. Loss of air (and thus compaction) may also occur during pumping. These investigators have demonstrated that the increase in flow resistance is directly proportional to the amount of compaction (air loss), which in turn depends directly on the air content of the mixture before pumping (they have observed that, irrespective of the initial air content, the final in-place air content of wet-mix shotcrete is always quite close to a basic value of approximately 4%). Figure 12 shows the effect of compaction during pumping and shoooting on the flow resistance.

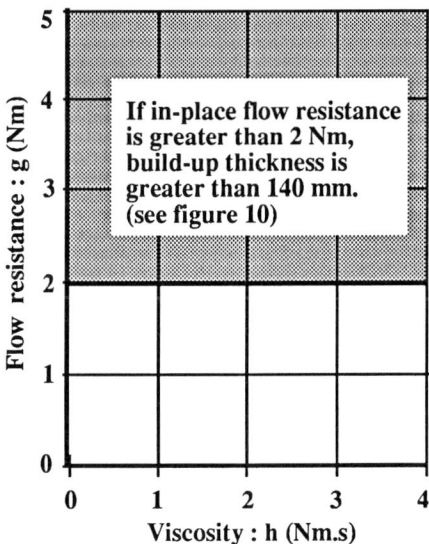

Fig. 11: Workability box for a minimum build-up thickness of 140 mm

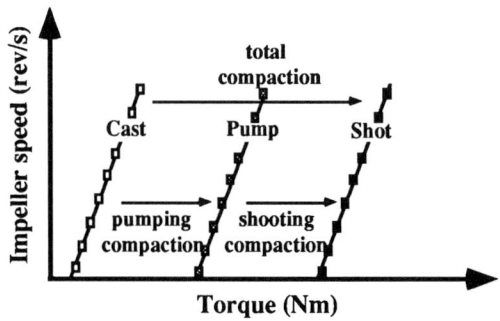

Fig. 12: Effect of compaction on flow resistance

3.3 Workability box

From the relationships presented in the previous Figures, it is possible to draw a workability box for the complete process of pumping and shooting wet-mix shotcrete. Figure 13 explains how such a box can be obtained taking into account both the pumpability requirements (Figure 13a) and the shootability requirements (Figure 13b), as well as the effect of compaction (lower part on Figure 13c). This workability box is valid only for a mixture pumped with the equipment used to generate the relationship in Figure 13a and for a selected minimal build-up thickness of 140 mm (Figure 13b). If the mixture undergoes compaction during shooting, the complete box extent toward lower limits of acceptable flow resistance before pumping.

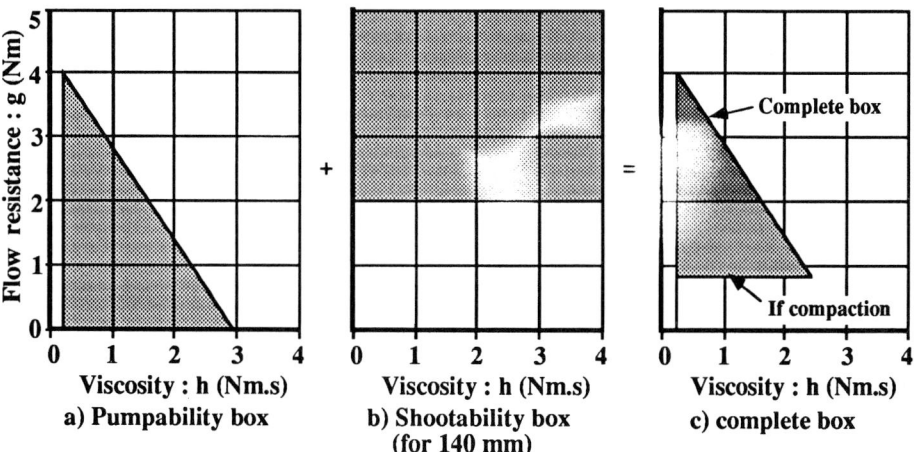

Fig. 13: Workability boxes for air-entrained wet-mix shotcrete.

4 Conclusion

With the recent studies performed to understand the fundamentals of wet-mix shotcrete, the workability box concept proposed by Tattersall can now be applied to this technique. Other parameters such as the fresh concrete ageing rate (FCAR) or the influence of thixotropy must also be studied to obtain a complete understanding of this placement method.

5 Acknowledgment

The authors are grateful to the Natural Sciences and Engineering Research Council (NSERC) of Canada (through its Network of Centers of Excellence program) for its financial support. Thanks are also due to Michel Pigeon who reviewed the manuscript.

6 References

1. Tattersall G. H., (1991), *Workability and Quality Control of Concrete*, Chapman & Hall, London, 262 p.

2. Beaupré, D., Mindess S. (1993) Compaction of Wet Shotcrete and its Effect on Rheological Properties, in *International Symposium on Sprayed Concrete*, Fagernes, Norway, October 22-26, pp. 167-181.

3. Hu, C. (1995) Rhéologie des bétons fluides (Rheology of Fluid Concretes), Ph. D. thesis, Laboratoire Central des Ponts et Chaussées, Paris, France, 201 p.

4. Scullion T., (1975), The Measurement of the Workability of Fresh Concrete, Masters Thesis, University of Sheffield, 1975.

5. Beaupré, D. (1994) Rheology of High Performance Shotcrete, Ph. D. thesis, Department of Civil Engineering, The University of British Columbia, Vancouver, Canada, 250 p.

6. Browne R.D. and Bamforth P.B., (1977), Test to Establish Concrete Pumpability, *Journal of American Concrete Institute*, May, pp. 193-207.

7. Beaupré, D., Mindess S., Morgan, D.R. (1993) Development of High Performance Shotcrete (Theoretical Considerations), Engineering Foundation Conference: *Shotcrete for Underground Support VI*, Niagara-on-the-Lake, Canada, May 2-6, pp. 1-8.

PART EIGHT

TEST METHODS AND PERFORMANCE

29 EVALUATION AND TESTING OF SPRAYED CONCRETE REPAIR MATERIALS

P. Lambert
Mott MacDonald, Manchester, UK
C.R. Ecob
Mott MacDonald, Croydon, UK

Abstract
The increasing use of spray applied materials for structural repair has highlighted the need for meaningful and reproducible testing regimes. While the manufacturers of pre-bagged materials provide a wide range of test data on which to base specification, it is not always apparent which attributes are most important and how the material as applied can be shown to be fit for purpose. This paper reviews aspects of performance testing for such materials as supplied by the manufacturer and as obtained on site. Results from a large repair programme are used to illustrate aspects of the testing and validation of repair materials in-situ.
Keywords: Repair, site testing, pull-off tests, interpretation of results.

1 Introduction

Spray applied cementitious materials are being increasingly specified for use in major reinforced concrete repair programmes. Spray application offers a number of advantages in the successful placement of repairs, especially on soffits and vertical faces. Large volumes of such materials have been used for reinstatement of carbonated or chloride contaminated cover, strengthening works and as overlays for cathodic protection systems.

The performance of any repair material is clearly critical to the long-term success of the remedial works in which they have been employed. While it should be relatively straightforward to identify and specify the appropriate performance characteristics required by a particular application and match these to available materials, the inevitable complications of site application may result in less than ideal performance.

Manufacturers of sprayed concrete repair materials now regularly provide a wide range of test results upon which to base an informed selection, although there is still much confusion regarding which parameters are important for specific applications. In-situ site testing of the applied materials is also fraught with difficulties and uncertainties, particularly with respect to demonstrating adequate adhesion to the substrate.

2 Performance Testing

A wide range of attributes are now routinely tested by the manufacturers of pre-bagged spray-applied repair materials and provided for the specifier or user on data sheets. The results are generally obtained from cores taken from test panels. Where appropriate, these can be verified with cores taken from test panels spayed on-site under representative conditions.

Selecting which aspects are of greatest significance with respect to any one application can be difficult while comparing the relative performance of alternative materials is often complicated by the non-standardisation of the test methods.

For most applications the performance attributes of greatest importance, in no particular order, are as follows:

- Compressive strength
- Rate of strength gain
- Modulus
- Thermal expansion
- Shrinkage
- Adhesion
- Gas and vapour transport characteristics
- Resistivity (particularly for cathodic protection applications)

The ultimate strength of these materials is generally far in excess of that required for the repair, it is often therefore more important that the rate of strength gain is appropriate for the planned work programme. Traditionally the high strength of repair mortars went hand in hand with high stiffness. A consequence of this was that the repairs tended to debond when subjected to flexure. Some more recently formulated materials have elastic moduli comparable with 'normal' grades of structural concrete.

Control of shrinkage and achieving good adhesion are essential in repair, particularly where relatively thin layers are used such as for cathodic protection overlays. It is likely that a properly applied spray repair material would provide significantly improved resistance to carbon dioxide and chloride ion ingress when compared to the parent material, further enhancing the repair characteristics.

As a result of the high density of spray applied repairs, together with the inclusion of pozzolanic and polymer modifiers, such materials usually exhibit relatively high electrical resistivity. In conventional repairs this can help to control the magnitude of corrosion currents but in cathodic protection applications where the flow of current is a requirement exceptionally high resistivities may be a disadvantage. The influence of resistivity on the operation of cathodic protection systems has been discussed elsewhere [1].

3 In-Situ Testing

While it is now relatively straight-forward to specify an appropriate spray-applied repair material based on manufacturer's data sheets, there remains the problem of demonstrating that the material as placed is of an appropriate quality. Full and proper specification, supervision and the use of experienced applicators will go a long way to ensuring that all aspects of handling, storage, surface preparation, application and curing are carried out in an appropriate manner to ensure optimum performance from the repair material.

Testing of the repair material as applied is generally realistically limited to close visual inspection, delamination surveys and pull-off tests. Detailed visual examination by an experienced person is probably the most cost-effective way of establishing the general quality of an application in addition to checking any aesthetic requirements. The main disadvantage is the lack of standardisation and reproducibility.

Delamination testing by means of a hammer survey is another valuable and relatively inexpensive method of establishing the quality of an application, particularly thin layer applications such as cathodic protection overlays. Again, experience is everything, from selection of a suitable hammer, to the interpretation of the response. Surveying a highly textured as-sprayed gunite can prove confusing to even an experienced operator and differences of opinion between the various parties on-site are all too common.

Clearly what is required is a relatively simple test that can be carried out to a recognised standard thus limiting the influence of the operator and variation in interpretation. Pull-off tests to BS 1881:Part 207 [2], appear to offer such a test. In addition to the credibility of a British Standard, there is supporting guidance from reputable and trusted sources such as the Concrete Society [3], and CIRIA [4] to assist in their interpretation.

4 Interpretation of Pull-Off Tests

While carrying out pull-off tests correctly and in a reproducible manner can in itself be difficult, the interpretation of the results is often also a source of confusion, despite the considerable assistance that is available, as referenced above.

It is common for repair specifications to require pull-off tests to demonstrate a minimum value, for example 1 MPa. This may in itself cause confusion if the nature of the failure is not also taken into account. To illustrate this, actual results from a concrete repair contract are given in the following table.

Table 1. Pull-Off Test Results (from large-scale soffit repair programme)

Failure Stress (MPa)	Failure Mode (Frequency)				TOTAL
	Glueline	Gunite	Interface	Substrate	
0	0	0	0	0	0
0.1	0	0	0	0	0
0.2	0	1	1	4	6
0.3	1	0	0	4	5
0.4	0	0	1	4	5
0.5	0	0	0	6	6
0.6	4	0	0	5	9
0.7	5	0	0	8	13
0.8	4	0	0	6	10
0.9	1	0	0	1	2
1.0	3	0	0	9	12
1.1	3	1	1	15	20
1.2	8	1	0	12	21
1.3	2	2	1	4	9
1.4	8	1	0	5	14
1.5	0	0	0	7	7
1.6	2	0	0	8	10
1.7	0	0	0	3	3
1.8	2	2	0	8	12
1.9	3	0	0	3	6
2.0	1	1	0	4	6
2.1	3	0	0	2	5
2.2	2	0	0	1	3
2.3	1	0	0	2	3
2.4	2	0	0	1	3
2.5	0	0	0	0	0
2.6	0	0	0	0	0
2.7	1	0	0	1	2
2.8	0	0	0	0	0
2.9	0	0	0	1	1
3.0	0	0	0	0	0
TOTAL	56	9	4	124	193

The pull-off test results shown in Table 1 were obtained from large scale soffit repairs totalling around 200m^3 and were carried out in accordance with BS 1881:Part 207:1992 having first cored through the gunite and into the parent concrete.

The results are fairly typical for gunite replacement of cover (ie. following removal of carbonated or delaminated concrete) and cathodic protection overlays.

In both these types of application the integrity of the bond between gunite and the original substrate is critical with regard to the long-term durability of the repair.

When quoting a minimum value for pull-off results it is usually intended that the value relates to the cohesive strength of the gunite (shown as 'Gunite' on the table) or the adhesive strength of the bond between gunite and substrate (shown as 'Interface'). It is generally acknowledged that a proportion of the tests will fail at the bond between the dolly and the gunite surface (shown as 'Glueline') but that the value obtained may be used as an indication of the lower limit of the strength of the repair.

Closer examination of the results in Table 1 highlights a common problem encountered when interpreting the results due to the high proportion of tests that fail cohesively within the parent concrete (shown as 'Substrate'). Modern pre-bagged gunite repair materials, when properly applied and cured, typically demonstrate tensile strengths in excess of the parent concrete resulting in a very high proportion of 'substrate' failures.

In the example given, nearly two thirds of the tests resulted in failure in the parent material. Failure in the original substrate is generally regarded as a pass irrespective of failure stress, although regular failure at low values could be an indication of a weak or poorly prepared substrate.

Of the one third of tests that did not fail in the substrate, the majority were associated with glueline failures. In this case, any tests with a value above the specified minimum may be considered to have passed while those below have not failed but require retesting.

Failure in the gunite itself or in the bond between the gunite and the substrate account for around 5% and 2% respectively. Based on a 1 MPa minimum pull-off strength, only 3 tests failed outright (around 1.5%) while 15 required retesting due to glueline failure (around 8%). These subsequently passed.

6 References

1. Concrete Society.(1989) *Technical Report No.36. Cathodic Protection of Reinforced Concrete.* The Concrete Society, London.
2. British Standard Institution.(1992) *Testing Concrete. Part 207. Recommendations for the assessment of concrete strength by near-to-surface tests.* BSI, London. BS1881:Part 207.
3. Concrete Society.(1991) *Technical Report No.38. Patch Repair of Reinforced Concrete. Model Specification and Method of Measurement.* The Concrete Society, London.
4. CIRIA.(1993) *Technical Note 139. Standard Tests for Repair Materials and Coating for Concrete, Part 1, Pull-off Tests.* CIRIA, London.

30 EVALUATION OF NEW AND OLD TECHNOLOGY IN DRY PROCESS SHOTCRETE

R.J. Liptak and D.J. Pinelle
Conproco Corp., Hooksett, USA

Abstract
During the late 1980's and into the 1990's, the concrete repair industry in the United States has seen the birth of various products and systems designed to place materials faster, better, and more economically. In keeping with this trend and customer demands, a test program was initiated to explore available performance enhancers and raw materials in the quest for an improved dry process shotcrete product. Additionally, it was hoped that the test program would assist in selecting raw materials as a guide for design of a product with an optimum balance of properties. It is the authors' intention to present this data as an avenue to communicate the much needed research findings of both new and old shotcrete technology to the industry.

The candidates chosen for this study included an unmodified control, as well as those including fibers, silica fume, polymer, accelerator, and superplasticizer. Additional specimens consisted of various combinations of the listed additives. Results were then compared to previous work done by the authors and others, as well as formulations that have been successful in repair mortars other than shotcrete.

Results indicate the use of a properly selected accelerator imparts the added ability to create architectural details with the in-place shotcrete. Another benefit realized with the addition of an accelerator was an increase of the initial tensile creep. The ability to create architectural details is an indication that the accelerator promotes initial tensile creep. Tensile creep in combination with other properties has recently been used to predict dimensional stability in repair mortars. Also, the use of superplasticizer in conjunction with silica fume enhanced dry-process shotcrete appeared to be critical in these formulations.

Keywords: Tensile creep, balance of properties, dimensional stability, performance enhancers, prepackaged dry process shotcrete, architectural details.

1 Introduction

As we approach the year 2000, the United States is experiencing change in the way materials and processes are used to repair deteriorated concrete. Performance enhanced hand applied mortars are no longer the staple of applicators' repair methods. When large quantities of repair material must be placed, more economic and logistical methods such as shotcrete, low pressure spray, and form and pump/pour mortars are now common for the restoration of concrete.

Regardless of application method, the biggest challenge faced today is the ability for the "cavity filler" to remain dimensionally stable. It was hoped that portions of a mathematical model previously used with hand-applied mortars could predict crack resistance of several shotcrete formulations. Technology utilized to reduce cracking gave sight to another feature -- a delayed or latent final set. This enabled the shotcrete to be shaved or carved to create designs such as waffle-slab configurations some time after shooting. It should be noted that the majority of this study is presented from actual test shootings in which various mix designs have been compared through observation and the most limited amount of testing possible, with in place, shotcrete.

2 Perspective

Prepackaged shotcrete materials are now and will be in the future, a cost effective means of placing repair materials for both large and small volume repairs. In the past the reputation of dry process shotcrete in the United States has suffered from inconsistencies in material (field mix) and application performance. It is for these reasons that prepackaged shotcrete has emerged in the same fashion as non-shrink grout. For example, non-shrink grout was at one time exclusively prepared as a field mix at the job sight. As we know today, it is almost unheard of to use anything but prepackaged grout for applications as complex as base plate grouting, keyways, etc. As a quest for dimensionally stable repair materials continues and becomes more complex, the presence of prepackaged shotcrete in some form or fashion will become more prominent.

3 Review of repair envelope and predicting dimensional stability

The repair of concrete is more complex than most realize. The repair envelope is a very active system (Figure 1) beginning immediately after placing fresh material into the cavity. When moisture is lost to the atmosphere, substrate, and hydration, shrinkage occurs. Since the repair material is restrained by its bond to the parent concrete, stress results when shrinkage occurs. The induced stress will produce one of the following situations as seen in Figure 2.

Fig. 1. An example of the repair envelope concept.

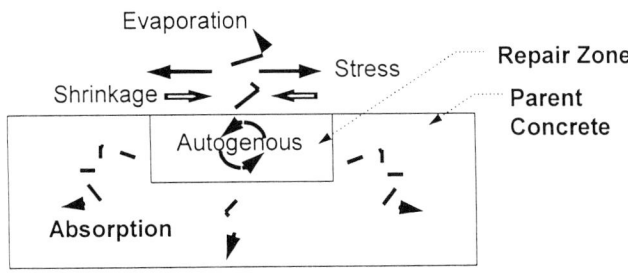

Fig. 2. A detailed diagram of what induced stress will produce in each of the following situations.

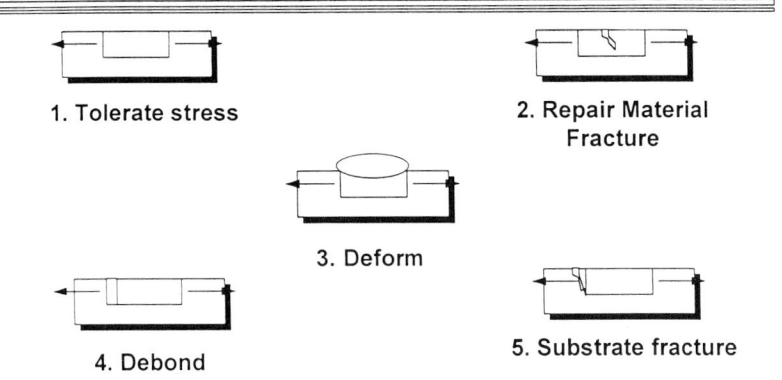

Since we want the repair material to remain crack free, it must be designed to resist tensile strain during the critical early stages of development. In previous studies a mathematical model was designed in order to predict a repair material's propensity to crack in both the short and long term stages of development. This study has been reported in other papers [1] [2], thus in order to keep within the scope of this study, we will only review the basic design parameters, which are as follows:

1. High tensile strength
Strength of material needs to exceed internal stress
2. Low modulus of elasticity
Movement/shrinkage causes less stress
3. High initial creep
Tensile deformation allows relief of stress
4. Low shrinkage
Less movement/ introduces less stress on curing

It has been found that a proper balance of the described criteria is essential for a repair material to remain dimensionally stable. Since shotcrete is essentially concrete/mortar, it was thought that portions of this model could be incorporated in a variety of mix designs.

4 Investigation and review of past studies and performance enhancers

Due to the vast amounts of similar technology available, the authors decided to design a shotcrete material that differed and/or optimized performance enhancers. The first step in this process was to review currently available performance enhancers as well as studies by other authorities and the authors' experiences.

4.1 Silica fume
It is well documented that silica fume has revolutionized the shotcrete industry as a performance enhancer [3] [4] [5]. Characteristics of this are as follows:

1. Rebound reduction
2. Increased buildup without accelerators
3. Resistance to environment and durability
4. Cohesiveness to resist washout in tidal areas
5. Increase in mechanical strengths
6. Partial cement replacement
7. Levels of 7-15 percent of cement are optimal
8. Increase in modulus of elasticity
9. Prone to surface crazing if overfinished or not cured

4.2 Superplasticizer
Studies and experiences by other authorities have indicated that superplasticizers are not useful with dry process shotcrete. The theory is that the reaction time for water, cement and silica fume is too short for effective water reduction by the time the material is applied [3].

4.3 Dispersant
A properly selected surfactant can be useful in combination with certain superplasticizers.

4.4 Accelerator
Accelerators were reviewed not necessarily for high early strength but for other desirable properties to be discussed later. Some research has found that accelerators may be detrimental to long term properties such as durability and compressive strength.

4.5 Fibers
The two most commonly used fibers are steel and polypropylene. Studies have shown that fibers reduce rebound, post crack resistance is lowered and impact resistance is improved [7].

4.6 Polymer
Although there are dozens of thermoplastic polymers available for use in shotcrete, acrylic and acrylic derivitives (styrene acrylic) have been the most widely used. Polymers are excellent for durability, internal curing, low modulus, and thin shotcrete applications. Thermoplastic polymers are available in two forms:

1. As an emulsion polymer.
2. As a dry powdered polymer.

Obviously, dry polymer is most commonly used for logistical and economic reasons. Some authorities have presented the theory that dry polymer does not disperse well in dry process shotcrete. The authors have done previous studies (unreported) with dry polymer that indeed indicate proper dispersion in both repair mortars and dry process shotcrete.

5 Discussion of selected mix designs

All formulations were prepared by continuous batch process for this test study. The description of the performance enhancers and ingredients are as follows:

- Portland Type I cement
- Silica sand
- Silica fume
- 1 cm. fiberglass
- Accelerator
- Superplasticizer
- Dispersant
- Vinyl versatate/Vinyl acetate polyvinyl alcohol stabilized dry polymer

Except for the field mix, all test candidates were constant in levels of sand, cement, and dispersant. A series of batches were then prepared incorporating the various performance enhancers one by one. Experimentation was also done with some combinations of the performance enhancers. A detailed list of each shotcrete can be seen in Table 1. The list below is a summary of the formulas experimented with.

1. Control
2. Silica fume
3. Silica fume without superplasticizer
4. Silica fume and accelerator
5. Silica fume and fiber
6. Silica fume, fibers, and accelerator
7. Polymer only
8. Field mix (prepared by contractor)

Table 1. Shotcrete mix designs

Additive	mix 1	mix 2	mix 3	mix 4	mix 5	mix 6	mix 7	mix 8
Portland Type I cement	1 part	1 part	1 part	1 part	1 part	1 part	1 part	1 part
Silica sand	3 parts	3 parts	3 parts	3 parts	3 parts	3 parts	3 parts	4 parts
Silica fume		X	X	X	X	X		
1cm. fiber					X	X		
Accel.				X		X		
Polymer							X	
Super-plasticier	X	X		X	X	X	X	
Dispers.	X	X	X	X	X	X	X	

6 Field applications and test variables

Date: 1994
Location: Madison, Maine, USA. Madison Electric Powerhouse
Conditions: 29-35 C, moderate wind
Equipment: Allentown GRH 600 double chamber gun
Test Panels: 76.2cm x 76.2 cm. x 10.16 cm. plywood panel
Hose: 3.175 cm., 1524 cm.
Air: 375 CFM
Curing: Continuous wet burlap
Application: All panels were shot by one experienced nozzleman. Panels were predampened with water, shot, then wet cured with burlap. Once final set took place, panels were screeded and then shaped if possible with steel trowels. Panels were then transported for laboratory analysis while still continuing to cure for three days.

7 Field observation

Overall, the best performing candidate was the combination of silica fume, fiber, and accelerator (mix 6). Rebound, slough, dust, screeding, and shaping/carving were superior. Final set was surprisingly quick, however, it did not affect shaping and carving characteristics. Fast set may have been attributable to high wind and temperature as this candidate was applied last.

The silica fume without superplasticizer shotcrete (3) showed evidence of significant sloughing, rebound was poor, and the material was difficult to shape/carve. It was noted that more water was needed to "wet" out material which led to sloughing.

The silica fume with superplasticizer shotcrete (2) showed significant improvements in the described properties as compared to the shotcrete without superplasticizer (3).

The polymer shotcrete (7) was easy to screed. However, this was due to severe set retardation. This shotcrete did not reach a final set for more than 24 hours. Because of the retarded set of sample 7, wet curing with burlap was removed shortly after shooting. Dust was unusually high.

The silica fume with fiber mix (5) was also very good in handling, however, the silica fume with accelerator (4) was unusually dusty and very slight sloughing was observed. Sloughing was still better as compared to silica fume without superplasticizer shotcrete (2).

The control shotcrete (1) displayed a substantial amount of rebound and dust. The field mix shotcrete (8) had less rebound and dust than the control shotcrete (1). This may have been attributed to high moisture in the sand.

8 Laboratory results

Compressive strength tests were run at 1, 7, and 28 days. When possible 3'x3' cubes were saw cut and tested in accordance with ASTM C-109 (modified for the larger cubes). The field mix and polymer formulation had very low strength development, as did the silica fume without superplasticizer as compared to the other silica fume shotcretes.

Flexural strengths were evaluated and tested according to ASTM C-348. All were fairly consistent after 28 days except for the field mix.

Cubes were also visually "core graded" for evidence of sand pockets. All were very good except for silica fume without accelerator and field mix.

Currently, there is no test method for measuring shrinkage with in place shotcrete. It was hoped that field performance of the best shotcrete candidate could be monitored for cracking. Future laboratory work is to include shrinkage and other tests that could not be performed on the in place shotcrete candidates. It should be noted, however, that rebound can change the composition of in place shotcrete and as a result, not allowing accurate comparison to laboratory mixes.

The best performing shotcrete from the field was the silica fume, fiber, accelerator combination. Considering the excellent field performance, panels from this sample were further studied for resistance to deicing salts. This shotcrete passed 50 cycles with no effect by using the ASTM C-672 method.

Table 2. Laboratory Results

Test	Related Info.	1	2	3	4	5	6	7	8
Compressive Strength ASTM C109 modified	Days 1	993	968	1113	1912	1706	1800	N/A	358
	7	2918	2006	1927	2956	2990	3504	1499	918
	28	6690	4847	3498	5172	5164	6141	4063	3501
Flexural Strength ASTM C348 (28 days)	"raw"	450	700	400	600	400	700	600	250
	"modulus of rupture"	675	1050	600	900	600	1050	900	375
Resistance to Deicing Salts ASTM C672							50 cycles at no effect		
Air Content (%)		16.0	5.1	7.7	6.0	2.3	4.0	6.4	6.0
Density		148	150	131	140	147	131	138	146
Vicat Set Time (mins.)	Initial	56	60	100	100	105	35	160+	63+
	Final	78	69	127	150	120	40	N/A	N/A

9 Field performance

1. Big Bethel Water Treatment Facility
   Big Bethel, Virginia, USA
   Digester Repair (September 1995)
2. James River Corporation
   Old Town, Maine, USA
   Resurfacing Brick (Spring 1995)
3. Echo Lake Association
   Winthrop, Maine, USA
   Dam Resurfacing and Repair (Summer 1995)

Each of the projects utilized a silica fume, fiber, and accelerator dry shotcrete mix and all were virtually crack free. The Big Bethel project had a small percentage of areas where the surface exhibited plastic shrinkage cracks. This was found to be due to large and rapid loss of surface water and a lack of wet curing.

10 Conclusion

Studies with shotcrete are often restricted because of logistics, economics, and time constraints. Since reproducing the mixing action of shotcrete in a laboratory is difficult, it was felt to be necessary to perform a majority of the research in the field. Field research, however, is expensive and includes many variables that cannot be controlled as accurately in the laboratory (eg., wind, temperature). Although laboratory analysis is more scientific, the authors felt that in addition to laboratory testing, actual field research was essential and more representative of the data needed for this study. The following conclusions were determined from this study:

- The silica fume, fiber, accelerator shotcrete had the best balance of properties as demonstrated in both field and laboratory.
- Superplasticizer was essential when used with silica fume to prevent sloughing from excess water demand. This indicated dispersion of superplasticizer, silica fume, cement, and water occurred.
- The vinyl acetate/vinyl versatate dry polymer demonstrated the ability to disperse by causing undesirable retardation of final set.
- Field performance results indicate silica fume, fiber, accelerator shotcrete remains dimensionally stable to date.
- The accelerator appears to impart the desired early tensile strength without affecting early tensile creep. When utilizing an accelerator, it was also possible to retain the ability to shape/shave architectural detail some time after shooting. The early tensile creep may also be helpful in relieving shrinkage induced stress.

- The theoretical increase in modulus of elasticity from silica fume does not appear to dramatically affect crack resistance in field performance.
- Use of silica fume is superior as a shotcrete performance enhancer when compared to the vinyl acetate/vinyl versatate. The combination of silica fume, fiber, and accelerator performed best in the field and laboratory.

11 Future work

- Perform analysis such as tensile strength, shrinkage, tensile creep, and modulus of elasticity in laboratory mixes for more accurate prediction of crack resistance.
- See if sand content can be increased.
- Use of metakaolin as replacement for silica fume.

12 References

1. Pinelle, D., Vaysbord, A., Emmons, P., Poston, R., Walkowicz, S. (1995) *Origin of Durability of Concrete Repairs.*
2. Pinelle, D. (1995) Curing stress in polymer modified repair mortars. *ASTM Cement, Concrete, and Aggregates,* Vol. 17, No. 2.
3. Morgan, D. (1988) Dry mix silica fume shotcrete in Western Canada. *Concrete International,* Vol. 10, No. 1.
4. Wolsiefer, J., and Morgan, D. (1993) Silica fume in shotcrete. *Concrete International,* Vol. 15, No. 4.
5. Liptak, R. (1993) Dry mix silica fume shotcrete. *Concrete Repair Bulletin.*
6. *Guide to Shotcrete, ACI 506, R 93* (ACI Committee 506).
7. Richardson, B. (1990) High volume polypropylene reinforcement for shotcrete. *Concrete Construction.*

13 Acknowledgments

Damaged Masonry Technologies, Maine, USA
Knowles Industrial Services Corporation, Maine, USA

31 STRENGTHENING OF CONCRETE BRIDGES USING REINFORCED SPRAYED CONCRETE

D. Pham-Thanh, S.R. Rigden, E. Burley
Dept of Civil Engineering, Queen Mary & Westfield College, University of London, UK

P. Quarton
Connaught Group Ltd, Gloucester, UK

Abstract
An extensive Department of Transport bridge assessment programme has revealed a significant number of highway bridges being deficient in their load carrying capacity to cope with the expected implementation of EC directives in the UK which would result in much heavier commercial vehicle axle loads being exerted on UK bridges. It has been recognised by many bridge owning authorities that there is an extreme urgency for their bridge stock to be strengthened to meet this requirement. Currently bridges can be strengthened by gluing steel plates onto the soffit of the bridge in the critically deficient moment regions to increase their moment capacity.

This paper describes an experimental investigation of an alternative technique of strengthening bridges using reinforced sprayed concrete as a strengthening layer on the soffit of concrete bridges. The technique is being simulated by strengthening previously cast reinforced concrete slabs by fixing an additional layer of mesh reinforcement below the slab soffit and then covering this with a layer of sprayed concrete. Eighteen slabs were cast, of which fourteen were statically load tested to failure and three cyclically. The variables considered in both static and cyclic loading were the size and the amount of the mesh reinforcement, the thickness and mix used for the sprayed concrete layer. Additionally, several experimental investigations were carried out to study how the horizontal shear capacity developed at the substrate/sprayed concrete interface is affected by five different types of surface preparations.

Keywords: Bridge, cyclic load, horizontal shear, shear connectors, shotcrete, slabs, sprayed concrete, strengthening.

1 Introduction

Research into the use of reinforced sprayed concrete for strengthening structural concrete members and in particular concrete bridge decks is sparse. The information available tends to relate to sprayed concrete as a material and shows a wide range of values. In view of this lack of data, it is hoped that the findings from the experimental investigation described in this paper will considerably improve knowledge in this area.

The success of this strengthening technique depends primarily on how effectively the applied sprayed concrete layer bonds to the concrete member being strengthened and how durable the sprayed concrete and the associated bond interface is when exposed to an aggressive environment, for instance freeze-thaw exposure.

The experimental investigation performed in the department of Civil Engineering of Queen Mary & Westfield college has demonstrated that at the stress levels imposed, the bond between the sprayed concrete layer and the substrate slab (or base slab) was extremely good without the aid of mechanical shear connectors and that composite action was maintained right up to total flexural failure of the increased depth slab. This finding was further reinforced in the experiments where the substrate/sprayed concrete interfaces were subjected to even higher stress levels by increasing the reinforcement in the sprayed concrete layer by up to four times that in the base slab and also by the quality data obtained from the extensive horizontal shear study experiments.

2 Experimental investigation - phase I

In this phase of the investigation, the performance of the technique was tested under static load only and the test was designed to measure the improvement in flexural capacities of the test slabs.

2.1 Base slab details

Nine slabs of 2.4 x 1.0 x 0.1m were cast in the laboratory. The concrete mix used in the slabs was a C35 designed concrete with 125mm slump supplied by Ready Mixed Concrete Limited. The aggregates size was 10mm maximum and the reinforcement was A193 mesh with deformed bars. After curing for 28 days, the slabs were placed on scaffolding at a height of two metres to simulate as near as possible the conditions under which a bridge soffit would be sprayed.

One slab was kept aside as an un-strengthened control slab and the underside of the remaining eight slabs were prepared by means of grit blasting. The extent of grit blasting was just sufficient to remove the concrete laitance so that the aggregate was exposed, followed which Hilti metal hangers were fixed to the underside of the slab and the mesh reinforcement rigidly secured to them. Timber spacer blocks of the correct size were used to set the mesh reinforcement at the pre-determined depth relative to the overall pre-determined thickness of the strengthened slab when sprayed. The eight slabs were then sprayed with reinforced sprayed concrete using the dry process. The whole strengthening process was carried out by Tarmac Structural Repairs. Figure 1. shows the reinforcement details.

In order to see whether the horizontal shear capacity at the substrate/sprayed concrete interface of the strengthened slabs would be enhanced, mechanical shear connectors were incorporated in three test slabs. Figure 2. shows the details of the shear connectors used.

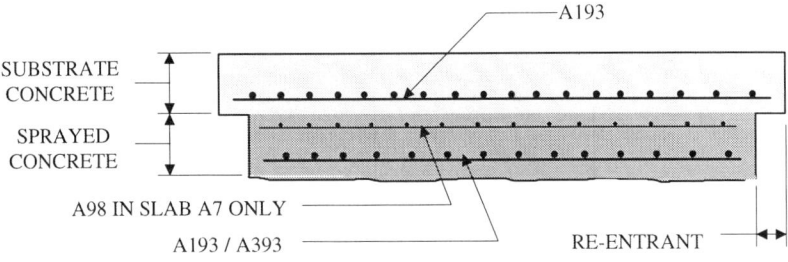

Figure 1. Typical test slab reinforcement details.

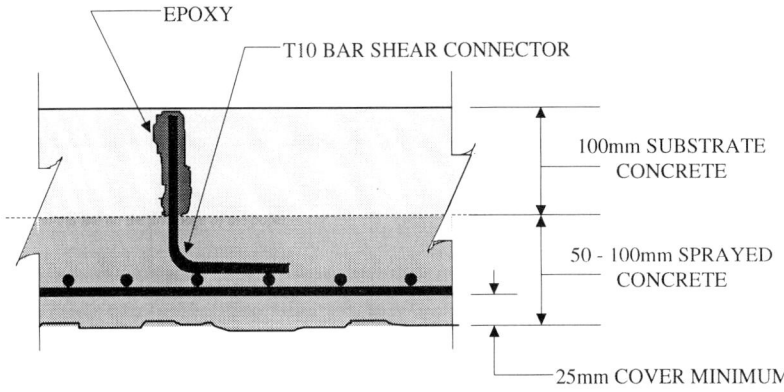

Figure 2. Shear connectors details.

2.2 Sprayed concrete details

Table 1. shows details of the a test slabs. The ratio of cement to sand was 1:3, with both cement and sand were supplied prepacked. Microsilica in slurry form was added to the water at 7 % by weight of water during the spraying process. The microsilica and equipment for accurate mixing with water were supplied by Elkem Materials Limited. The prepacked sand contained angular aggregates ranging up to 5mm in diameter.

Table 1. Test slabs details (all mesh reinforcement are deformed bars except for A98).

Slab	Sprayed concrete thickness (mm)	Reinforcement in concrete	Shear connectors
Control	-	-	-
A1	75	A193	-
A2	75	A193	-
A3	75	A193	YES
A4	75	A393	-
A5	75	A393	YES
A6	50	A193	-
A7	100	A98 & A193	-
A8	100	A193	YES

Using the same concrete mix as that used in the test slabs, the same equipment and the same crew, a total of six pre-construction test panels measuring 600 x 600 x 100mm were sprayed, to supply specimens for material testing and to provide an indication of the quality to be expected in the test slabs. Two of these panels were sprayed horizontally and four in the overhead position. Comparisons were made to see if the spraying direction affected the quality of the sprayed concrete used in the test slabs. On completion of the spraying process, the sprayed concrete was sprayed with a proprietary curing membrane and left to cure for 28 days.

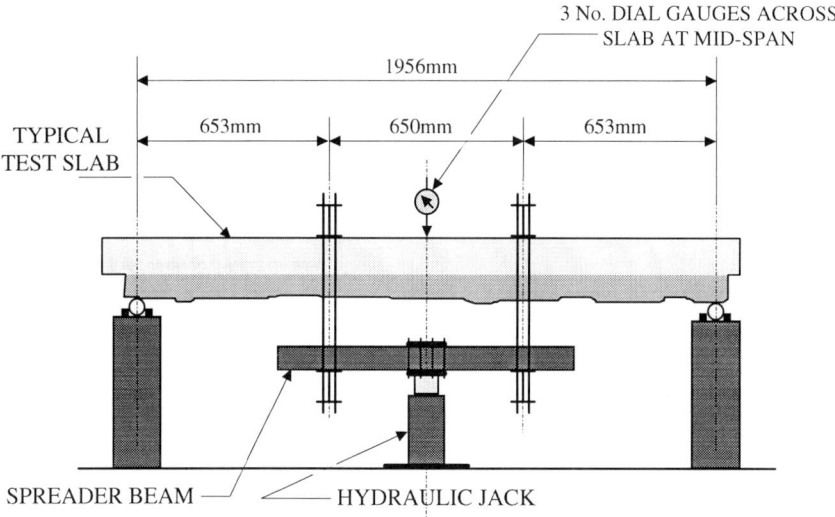

Figure 3. Load arrangement.

2.3 Experimental procedures

After 28 days of curing to the sprayed concrete, all the tests slabs were statically load tested to failure, using the experimental set up shown in Figure 3.

These slabs were simply supported over a span of 1956mm with the support bearings placed beneath the new sprayed layer of concrete and loaded at third points. The load was applied in 5KN increments up to failure. The deflections of the slabs were recorded by three dial gauges positioned across the slabs at mid-span.

3 Experimental results - phase I

3.1 Mechanical tests

Table 2. shows the mechanical test results obtained from specimens extracted from both the load tested slabs and the pre-construction panels.

Table 2. Mechanical test results phase I.

Test	Age	Specimen size (mm)	No. of specimens	Overhead/Horizontal sprayed	Strength/Coefficient
Compressive	40 day core	$\varnothing = 74$, L = 75.87 L/$\varnothing$ = 1.03	4	Overhead	35.0 N/mm^2
	40 day estimated cube strength	-	-	-	32.5 N/mm^2
	40 day core	$\varnothing = 74.12$, L = 81.07 L/$\varnothing$ = 1.09	4	Horizontal	35.9 N/mm^2
	40 day estimated cube strength	-	-	-	34.2 N/mm^2
	84 day core	$\varnothing = 50.81$, L = 55.52 L/$\varnothing$ = 1.09	2	Horizontal	41.4 N/mm^2
	84 day estimated cube strength	-	-	-	39.6 N/mm^2
Tensile splitting cores drilled parallel to direction of spraying	40 day core	$\varnothing = 99.78$, L = 103 L/$\varnothing$ = 1.03	3	Horizontal	3.6 N/mm^2
	40 day cube	98 x 98 x 98	4	Horizontal	4.9 N/mm^2
	84 day core	$\varnothing = 74.41$, L = 73.91 L/$\varnothing$ = 0.99	4	Overhead	6.2 N/mm^2
Tensile splitting loaded parallel to longitudinal plane of layering	84 day core	$\varnothing = 74.18$, L = 75.79 L/$\varnothing$ = 1.02	4	Horizontal	5.0 N/mm^2
Static modulus of Elasticity	84 day	32.54 x 31.62 x 96.00	1	Horizontal	24.1 N/mm^2
Pull - off	84 day	$\varnothing = 50$	6	Overhead	1.1 N/mm^2
Slant shear bond	84 day	57.75 x 55.77 x 150	1	-	37.6 N/mm^2
Coefficient of thermal expansion	70 day	-	-	-	1.26 x 10^{-5}

3.2 Slant shear test results

The slant shear test is one of the most informative tests for assessing the success of the technique of reinforced sprayed concrete strengthening. It was found that on failure of the test specimen the fracture lines were distributed uniformly and symmetrically relative to the longitudinal axis of the specimen and the fracture lines crossing the slanted interface in the direction of the longitudinal axis of the specimen, a mode of failure that would be expected if a similar test specimen was fabricated monolithically. It therefore appears from the results of this test that grit blasting as a means of surface of preparation was more than adequate, as extremely good bond was maintained at the substrate/sprayed concrete interface.

3.3 Petrographic report

Samples of sprayed concrete were extracted from the tested slabs and the pre-construction test panels and transported to Geomaterials Research Services Limited for material analysis and for chloride ingression measurement.

Material analysis: The analysis conducted was very comprehensive and gave the following conclusions:
1. The supplied sprayed concrete repair material contains a siliceous aggregate ranging up to 5mm in diameter, in a medium grey Portland cement containing microsilica.
2. The majority of the paste has very low porosity and contains very few micro cracks.
3. The paste contains numerous voids. The voids show some tendency to cluster around the margins of fine aggregate pieces and may form discontinuous and continuous cavities which are orientated parallel to the internal and external surfaces.
4. The majority of the material is strong and robust but there are patches of weak and friable and highly porous concrete below the reinforcement rods. This poorer quality material is considered to be the result of interference from the rods with the spraying of the concrete.

Chloride ingress: This analysis arrived at the following useful conclusions.
The diffusion coefficient found is similar to but less than that normally encountered in concrete at about 5×10^{-12} m/sec. The profile is very steep and the surface concentration is considered to be less than might be expected from an ambient chloride concentration of 5 molar at a temperature of 38°C. The diffusion coefficient tends to increase slightly with depth into the concrete and to decrease slightly with time. Considering the mean of three analyses shows that this concrete gives a results almost identical with that expected for high quality concrete. The variation found in the outer 3mm reflects variation in aggregate proportion which is more erratic than that of the bulk of the material. The inner data are more systematic and suggest a steady rate of penetration of the chloride front (producing 0.1% chloride, 0.08% above background) of less than about 0.02 mm/day. This results in very close to that expected for high quality concrete.

3.4 Load test results

Figure 4. shows the load versus average mid-span deflection of all test slabs and the control slab. Table 3. shows the theoretical and experimental moments for all slabs at failure. The theoretical flexural failure moments were determined assuming that the

Table 3. Flexural moments.

Slab	Sprayed concrete thickness (mm)	Reinforcement in sprayed concrete	Shear connectors	Theoretical M_{cap1} (KNm)	Experimental failure jack load (KN)	Experimental M_{cap2} (KNm)	Ratio of $\frac{M_{cap2}}{M_{cap1}}$
Control	-	-	-	6.1	31.5	8.8	1.44
A5	75	A393	YES	28.0	117.0	38.2	1.36
A4	75	A393	-	28.0	104.0	34.0	1.21
A7	100	A98 & A193	-	24.4	117.0	38.2	1.57
A8	100	A193	YES	18.8	90.0	29.4	1.56
A1	75	A193	-	18.0	85.5	28.0	1.56
A3	75	A193	YES	18.0	81.0	26.4	1.47
A2	75	A193	-	18.0	76.5	25.0	1.39
A6	50	A193	-	15.5	81.0	26.4	1.70

Notes: The followings are applicable to the data tabulated above,
1. Factors γ_m of 1.5 for f_{cu} and 1.15 for f_y are included in calculating the theoretical values.
2. A98 in slab A7 assisted structurally in carrying load and was considered in the calculation.
3. A5 failed in a ductile manner and the slab remained intact therefore the actual effective depth could not be determined, but was assumed to be similar to slab A4.
4. All theoretical values are determined based on the actual effective depths of reinforcement layers as measured manually from the broken faces of the tested slabs.

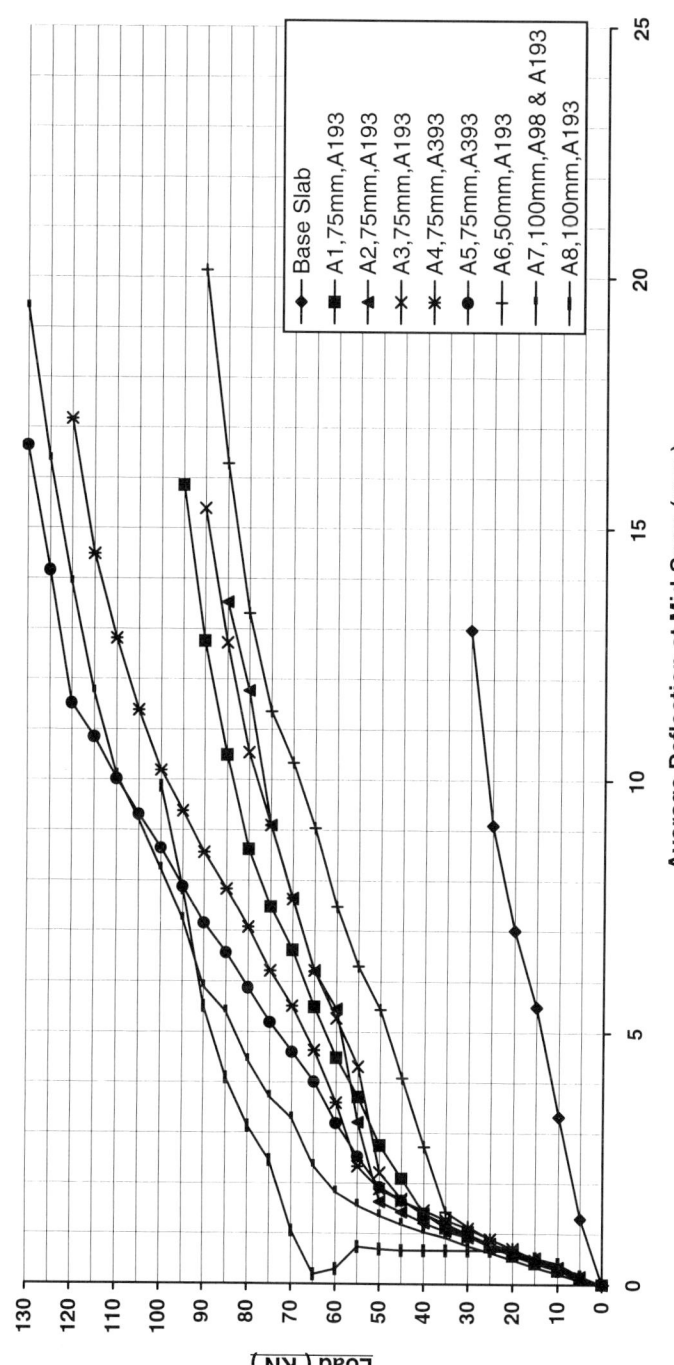

Figure 4. Load versus average deflection for all slabs.

reinforcement in both the sprayed concrete and the substrate concrete had yielded and that there was extremely good bond at the substrate/sprayed concrete interface.

4 Experimental investigation - phase II

In this phase of the investigation the performance of the technique was tested under static and cyclic loading and two additional experiments were conducted to study the horizontal shear capacity at the substrate/sprayed concrete interface. The results obtained from both shear tests will be compared, taking into account the slant shear test result from phase I.

In some slabs the encased reinforcement in the sprayed concrete layer is three to four times that in the substrate concrete slab. The purpose of this increase is to subject the substrate/sprayed concrete interface to a much higher shear stress level.

4.1 Reinforced Sprayed Concrete Curtailment

It was realised during the course of the research that for this technique of strengthening to be used on an existing concrete bridge structure, the reinforced sprayed concrete layer could extend to near the points of support and not beyond them. The original bridge bearings would still be used and the critical vertical shear capacity at the supports would not be improved. This was not considered to cause a major problem as our understanding is that the majority of slab bridges failing the Department of Transport bridge assessment programme do so due to lack of flexural capacity not shear capacity.

However some concern existed over the high longitudinal shear stress concentration that might be caused by the re-entrant corner at each end of the sprayed concrete layer which might cause peeling off of this layer at the ends.

In order to investigate this problem one slab with a large amount of reinforcement in the sprayed concrete layer was prepared to simulate the above end conditions by vertically cutting through the new sprayed layer to the depth of the substrate concrete in the span side of the support, see Figures 5. and 6. This slab was then subjected to static loading up to 9% beyond the designed failure load, followed by cyclic loading as discussed below.

4.2 Load test procedures

A total of nine strengthened slabs were fabricated; six were tested statically and three cyclically. In cyclic loading, the load range was between 40% and 50% of the design ultimate failure load, simulating the load range that a typical mid-span trunk road bridge would be subjected to with the passage of heavy commercial vehicles. Each slab was cyclically load tested in this load range for one million cycles at a frequency of 0.5 cycle per second. The structural integrity of each slab was then assessed before imposing a larger stress range between 30% and 70% of the static failure load to take it to failure. The maximum and minimum loads and their corresponding deflections were fed into a multi-channel computer controlled data acquisition system so that a full history of cyclic behaviour was obtained.

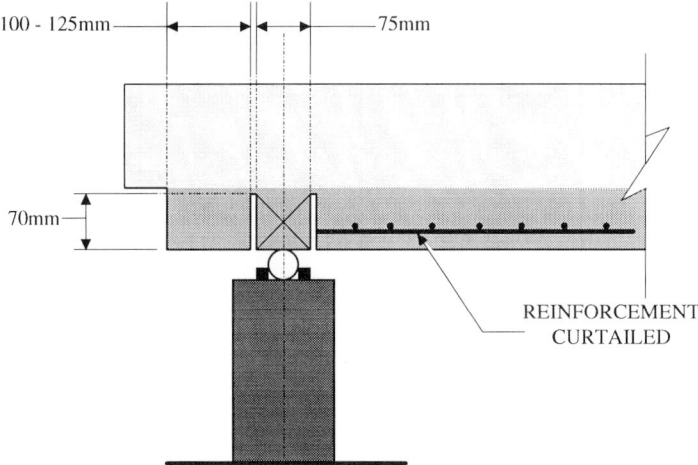

Figure 5. Dimensions of curtailment.

Figure 6. Reinforced sprayed concrete curtailment.

4.3 Horizontal shear study - Double shear test

One of the most important factors required for this technique of strengthening to be effective is that the composite action is maintained right up to failure. How well composite action is maintained depends on how well the additional sprayed concrete layer is bonded to the existing concrete member being strengthened i.e. the required horizontal shear capacity at the interface. To investigate this problem in greater details, ten blocks of concrete were sprayed on both sides to form a double shear test specimen. The sandwiched part was cast with the same concrete as that in the test slabs and the two sprayed parts were sprayed with the four sprayed concrete mixes used in the test slabs. Additionally, five different surface finishes were used, namely grit blasted; left as cast; scabbled; left as cast with reinforcement shear connectors and grit blasted with reinforcement shear connectors. A typical shear block is shown in Figure 7.

Under load these blocks simulate the shearing action on the test slabs and therefore provide a direct measure of the shear performance at the substrate/sprayed concrete interface.

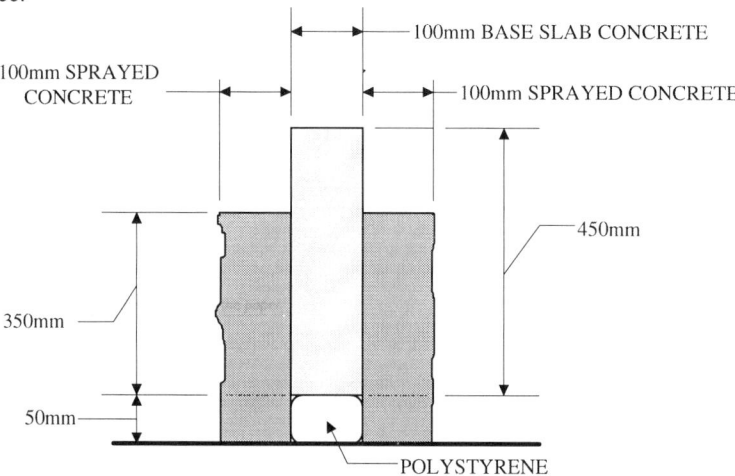

Figure 7. Details of a typical shear block (500mm overall length).

4.4 Horizontal shear study - Direct shear test

The authors developed this test so that it could be an economic but effective means of measuring the horizontal shear capacity at the substrate/sprayed concrete interface. It is simple to carry out and requires minimum labour from the initial stage of coring specimens to finally shear testing them. It therefore offers a useful and rapid field verification test.

During this test, the composite concrete specimen with its sprayed portion (or its substrate portion) would be held rigidly in position with the other portion being sheared along the plane of the substrate/sprayed interface. Figures 8. and 9. show the prepared test specimens and the shear loading apparatus.

5 Experimental results - phase II

Cyclic load testing of the slabs has now been completed but, the data is still being analysed at the time of writing this paper and therefore is not presented here. However, mechanical tests were carried out using specimens of sprayed concrete collected from the pre-construction test panels sprayed during phase II and the data is as shown in Table 4.

Strengthening of concrete bridges 291

Figure 8. Prepared direct shear specimens.

Figure 9. A direct shear specimen being tested.

Table 4. Mechanical test results, phase II

Mix	Sample size (mm)	Age (days)	Compressive		Tensile splitting	
			Overhead sprayed	Horizontally sprayed	Overhead sprayed	Horizontally sprayed
Normal	75Ø	47	29.1	26.1	3.7	3.5
Sikacem 133	75Ø	47	-	35.4	-	4.71
Steelfiber	75Ø	47	40.6	48.6	7.73	7.34
Polyfiber	75Ø	47	-	33.1	-	4.19

Notes:
- Normal = Sand + cement + microsilica.
- Sikacem 133 = Pre-pecked Sikacem 133 concrete + microsilica.
- Steelfiber = Pre-packed steel fiber concrete + microsilica.
- Polyfiber = Pre-packed polypropylene fiber concrete + microsilica.

5.1 Horizontal shear capacity - comparison

It is likely that if this technique of strengthening is adopted on a real structure, the surface preparation of the substrate concrete would be grit blasting and so an indication of the adequacy of this type of surface preparation under laboratory conditions would be useful. Figure 10. shows from the double shear test that for normal mix sprayed concrete with grit blasted surface preparation, a significant surface area of the substrate concrete was left on the sprayed concrete parts after failure and the shear capacity of this specimen was 2.14N/mm^2, much greater than that from CP110: Part 1: 1972 cl 5.4.3.4 as shown in Table 5.

Figure 10. A tested double shear specimen.

Table 5. Double shear test results.

Type of preparation	Grit	Scabbled	shear connectors only	Left as cast
Bond area (mm^2)	300x500	300x500	300x500	300x500
No. of specimens	1	1	1	1
Age (days)	100	100	100	100
Shear stress at failure (N/mm^2)				
Normal	2.14	1.69	2.00	0.77
Sikacem 133	5.93	-	-	-
Steelfiber	-	-	-	-
Polyfiber	3.37	-	-	-
CP110: cl 5.4.3.4	0.6	-	0.4	-

The results from the direct shear test (see Table 6) again indicates that extremely good bond had been achieved at the substrate/sprayed concrete interface with the shear capacity of the normal specimens being 61% that of the monolithic specimen and much higher than that from CP110 in Table 5.

The difference in the shear capacity of the normal specimens from the monolithic specimens could be explained by the relatively lower quality base concrete at the soffit compared to that in the middle of the base slab where the shear plane of the monolithic specimens was imposed. Another possible explanation for the difference could be that, to some extent, shear stress was set up in the base concrete near the interface due to the shrinkage of the sprayed concrete layer.

Table 6. Direct shear test results.

	Normal	Sikacem 133	Steelfiber	Polyfiber	Monolithic
• No. of cores	3	4	3	4	4
• Age (months)	19	19	19	19	19
• Average failure shear stress (N/mm^2)	2.44	3.59	2.84	2.77	3.98
• As % of monolithic	61.31	90.20	71.36	69.60	100.00

As described before in section 3.2, on the failure of the slant shear test specimen extracted and tested during phase I work, the fracture lines crossed the slanted interface in the direction parallel to the longitudinal axis of the specimen. A similar specimen extracted at the same time but now tested at a much older age of 21months showed a clear failure plane parallel to the slanted interface at approximately 7mm on the substrate concrete side with the recorded compressive strength of 43.7N/mm^2. This is 92% of that for the similar monolithic specimen extracted from the tested substrate concrete slab, see Table 7. Clearly very good bond at the interface was achieved

Table 7. Slant shear test results.

Slant shear bond strength	
Composite specimen	Monolithic specimen
• Specimen extracted from a tested slab. • Normal mix sprayed concrete. • Dimensions 55 x 55 x 150 mm. • 1 No. specimen at 84day age, 37.6 N/mm^2. • 1 No. specimen at 21month age, 43.7 N/mm^2.	• Specimens extracted from a tested slab. • All substrate concrete. • Dimensions 55 x 55 x 150 mm. • 4 No. specimens at 19 month age, 47.1 N/mm^2.

6 Discussion of results

6.1 Phase I

1. A significant increase in strength above that of the base slab, ranging from 175% to 300%, is seen for all the strengthened slabs tested.
2. No de-bonding of the substrate/sprayed concrete interface was observed in any test slabs.
3. Flexural cracks propagated across the substrate/sprayed concrete interface.
4. Comparison of similar slabs with and without shear connectors in Table 3, (see A4 & A5 and A1 & A3) shows that the inclusion of shear connectors does not cause any increase in strength. Comparison of A7 & A8 is more difficult since it was thought necessary to include a subsidiary mesh near the interface with the thick layer of sprayed concrete when shear connectors were not present and this extra reinforcement is the main reason for the increase in strength.

    Due to variations in slab thickness and reinforcement position, a more appropriate comparison is perhaps the ratio between experimental and theoretical values. This ratio is seen to vary between 1.21 and 1.70 with the average value for the strengthened slabs being the same as that for the base slab. Comparison of the results for the effects of shear connectors again confirms that no enhancement in strength is achieved by their use.

    The above comparison also shows that current composite theory can be used to predict the behaviour with a reasonable degree of accuracy. It appears therefore that at the stress levels imposed on the test slabs, shear connectors can be omitted without affecting the structural adequacy of the composite action of the slabs. This finding is also reflected in the slant shear test where no bond failure occurred and in the pull-off test where no de-bonding at the interface was found.

5. In the pull-off test, all cores broke in the sprayed concrete portion, indicating lower strength in this layer in comparison to the substrate concrete; this is also the case when comparing the compressive strength of the sprayed concrete to that of the substrate concrete.
6. The results show that the compressive strength of the microsilica sprayed concrete increased by 18% with ageing from 40 to 84 days. However, it should be noted that the 40 day tests were carried out on 75mm diameter cores and the 84 days tests on 50mm diameter cores.

    The compressive strength of the sprayed concrete does not seem to be affected by the direction of spraying with only 25% reduction in strength for overhead spraying compared with vertical spraying. However, voids were to be found

behind the reinforcement in some areas of the overhead sprayed slabs which could cause concern if action is not taken to limit this during the spraying operation.
7. The tensile splitting strength of the microsilica sprayed concrete seems to lie within the same range as that measured in the pull-off tests. However, this latter test should not generally be used for measuring the tensile strength of concrete.
8. It was observed from the slant shear specimen that there was a distinct dark layer at the interface, approximately 5mm thick. It is possible that this layer has a higher concentration of microsilica than the 7% dosage intended for the sprayed concrete. This possibility and its effect on the long term performance of the concrete is being investigated further.

6.2 Phase II

1. Using BS8110: Part 2 only for a general comparison, it can be seen from Tables 5. and 6. that the bond of the normal mix sprayed concrete to its grit blast prepared substrate is extremely good. However, it could be argued that this result may not be truly representative of the actual horizontal shear capacity developed in the test slabs, because of the horizontal spraying direction of these shear blocks as opposed to the overhead spraying direction in the test slabs. The results given in Table 4. for compressive strength and tensile splitting strength, show that there is no difference in the quality of the sprayed concrete whether sprayed overhead or horizontally.
2. Although data analysis is still to be carried out, careful examination of the six slabs tested statically and three slabs cyclically in this phase showed that there was extremely good bond between the substrate concrete slab and the sprayed concrete layer so that full composite action was maintained right up to failure without shear connectors incorporated and the stress levels imposed in this phase were much higher than in phase I. These higher stress levels are due to a significant increase in reinforcement encased in the sprayed concrete layer of three to four times the reinforcement in the substrate slab.
3. The results from the static and cyclic loading on the curtailed test slab showed that this slab failed in flexure at mid-span without de-bonding or cracks initiating at the re-entrant corners in front of the supports i.e. where the cut lines were made to the reinforced sprayed concrete layer. It may therefore be concluded that spraying and reinforcing a concrete layer on the bridge soffit, covering the area between the points of contraflexure would be an extremely effective strengthening technique.
4. Comparison of the results from slant shear, double shear and direct shear is given in section 5.1.

7 Acknowledgements

The work described in this paper is part of a research programme supported by the Engineering and Physical Science Research Council. A very major contribution to the work was made by Tarmac Structural Repairs in carrying out all the concrete spraying, supplying materials and advising on the test programme. The contribution of Elkem Materials Limited and Ready Mixed Concrete Limited who also supplied materials and equipment is gratefully acknowledged.

Author Index

Armelin, H.S.   243
Armstrong, K.   149
Austin, S.A.   107, 157

Banthia, N.   243
Barnes, R.A.   188
Barrett, C.   231
Beaupre, D.   3, 252
Boucheret, J.M.   166
Budelmann, H.   117
Burley, E.   280

Cabrera, J.G.   8
Choo, B.S.   81
Clay, R.B.   209

Dobson, R.B.   173
Dufour, J.-F.   3
Dykes, K.   49

Ecob, C.R.   265
Eisenhut, Th.   117

Farrell, M.B.   56
Figueiredo, A.D.   99
Ford, J.H.   181

Garshol, K.F.   26
Glassgold, I.L.   217
Gong, N.G.   81

Hackman, L.E.   56
Helene, P.R.L.   99

Iorns, M.E.   63

Jones, P.A.   107
Lambert, P.   265
Lamontagne, A.   3
Liptak, R.J.   270

Mai, D.   39
Mangat, P.S.   196
Manning, R.   70
Mays, G.C.   188
Mindess, S.   252
Morgan, D.R.   127, 243

O'Flaherty, F.J.   196

Peaston, C.H.   81
Pham-Thanh, D.   280
Pigeon, M.   3
Pinelle, D.J.   270

Quarton, P.   149, 280

Rich, L.D.   127
Rigden, S.R.   280
Robins, P.J.   107, 157
Ruffert, G.   226

Scott, M.J.   139
Seymour, J.   157

Takacs, A.P.   209
Turner, N.J.   157

Warner, J.   89
Woolley, G.R.   8, 231

Keyword index

This index is based on the keywords assigned to the papers. The numbers are the first pages of the relevant papers.

Admixtures 26
Age 99
Aggregate velocity 243
Air-entraining 181
Alkali-silica reaction 8
Alkalifree 39
Amorphous metallic fibres 166
Application technique 231
Architectural details 270

Balance of properties 270
Bond 81
Bridge 280
  repair 196
Build-up thickness 252

Cathodic protection 49
Cementitious materials 188
Certification scheme 231
Chloride ingress 149
Compaction 252
Composition 8
Compressive strength 3
Concrete 149
Contract 209
Corrosion 49, 149, 226
Creep 181
Curing 181
Cyclic load 280

Delamination 181
Density 8
Desalination 49, 149
Designer confidence 70
Dimensional stability 270
Dome 56
Dry
  shotcrete 117
  spraying 39
  -mix shotcrete 3
  -process 217

Durability 166

Ecology 39
Economy 26

Ferrocement 63
Fiber velocity 243
Fibre 127
  reinforcement 166
Fibres 26, 70, 139
Finite element analysis 81
Flexural strength 107
Floating slipforms 63
Flow resistance 252
Fly ash 8

Gunite 89

Highway structures 173
Horizontal shear 280
Humid aggregates 117

Interpretation of results 265

Laminating 63

Maintenance refurbishment 173
Marine 149
Masonry arches 81
Material properties 196
Materials 231
MEXE method 81
Mines 139
Mix design 70, 166
Modelling 107

NATM 70, 209
Non-toxic 39
Nozzle type 243
Nozzlemen 231

Offshore structures 63
Operatives 231
Ovoid-shaped sewage systems 166

Patch repairs 188
Performance 196
  enhancers 270
Permeability 8
Polymer-modified 181
Polyolefin 127
Post-crack behaviour 166
Pot-man 231
Powder
  accelerator 39
  air-entraining admixture 3
Prepackaged dry process shotcrete 270
Primary/secondary linings 70
Programmable dosing 117
Pull-off tests 265
Pull-out tests 107
Pumpability 252

Quality 26

Rapid hardening cement 117
Realkalisation 49
Rebound 8, 243
Redistribution of stress 196
Rehabilitation 226
Reinforced
  concrete 188
  shotcrete 89
Reinforcement 56, 149
Renovation 166
Repair 49, 157, 173, 265
  materials 196
Repairs 149
Rheology 157, 252
Rock cutting 139
Rockbolts 209

Safety 26
Scaling 3, 181
Shear connectors 280
Shootability 252
Shotcrete 26, 39, 56, 63, 99, 127, 209, 217, 252, 280
  application 89
  reinforcement 89
Shrinkage 181
Site testing 265
Slabs 280
Spacing factor 3
Specification 209, 217
Spray
  applied repairs 188
  cement 117
Sprayed
  concrete 8, 81, 89, 107, 139, 149, 196, 209, 231, 280
  mortar 173
Spraying process 157
Standards 217, 226
Steel
  fiber 56
  fibers 99, 107
Strength 8, 99
  increase 39
Strengthening 81, 280
Structural
  behaviour 188
  shotcrete 89

Tensile creep 270
Testing 49, 209, 217, 231
Tidal 149
Toughness 99, 127
Tunnel 209
Tunnels 139

Viscosity 252

Weighing and mixing installation 117
Wet
  mix 157
  process 107, 157
  sprayed mortars 166
  spraying 39
  -mix 252
  -process 217
Workability 166
  box 252